Geometry and Topology

Geometry provides a whole range of views on the universe, serving as the inspiration, technical toolkit and ultimate goal for many branches of mathematics and physics. This book introduces the ideas of geometry, and includes a generous supply of simple explanations and examples. The treatment emphasises coordinate systems and the coordinate changes that generate symmetries. The discussion moves from Euclidean to non-Euclidean geometries, including spherical and hyperbolic geometry, and then on to affine and projective linear geometries. Group theory is introduced to treat geometric symmetries, leading to the unification of geometry and group theory in the Erlangen program. An introduction to basic topology follows, with the Möbius strip, the Klein bottle and the surface with g handles exemplifying quotient topologies and the homeomorphism problem. Topology combines with group theory to yield the geometry of transformation groups, having applications to relativity theory and quantum mechanics. A final chapter features historical discussions and indications for further reading. While the book requires minimal prerequisites, it provides a first glimpse of many research topics in modern algebra, geometry and theoretical physics.

The book is based on many years' teaching experience, and is thoroughly class tested. There are copious illustrations, and each chapter ends with a wide supply of exercises. Further teaching material is available for teachers via the web, including assignable problem sheets with solutions.

MILES REID is a Professor of Mathematics at the Mathematics Institute, University of Warwick

BALÁZS SZENDRŐI is a Faculty Lecturer in the Mathematical Institute, University of Oxford, and Martin Powell Fellow in Pure Mathematics at St Peter's College, Oxford

Geometry and Topology

Miles Reid
Mathematics Institute, University of Warwick, Coventry CV4 7AL, UK

Balázs Szendrői
Mathematical Institute, University of Oxford,
24–29 St Giles, Oxford OX1 3LB, UK

CAMBRIDGE
UNIVERSITY PRESS

CAMBRIDGE UNIVERSITY PRESS
Cambridge, New York, Melbourne, Madrid, Cape Town, Singapore, São Paulo

Cambridge University Press
The Edinburgh Building, Cambridge CB2 2RU, UK

Published in the United States of America by Cambridge University Press, New York

www.cambridge.org
Information on this title: www.cambridge.org/9780521848893

First published 2005

Printed in the United Kingdom at the University Press, Cambridge

A catalogue record for this publication is available from the British Library

ISBN-13   978-0-521-84889-3 hardback
ISBN-10   0-521-84889-X hardback

ISBN-13   978-0-521-61325-5 paperback
ISBN-10   0-521-61325-6 paperback

Contents

List of figures		*page* x
Preface		xiii
1	**Euclidean geometry**	1
	1.1 The metric on $\mathbb{R}^n$	1
	1.2 Lines and collinearity in $\mathbb{R}^n$	3
	1.3 Euclidean space $\mathbb{E}^n$	4
	1.4 Digression: shortest distance	4
	1.5 Angles	5
	1.6 Motions	6
	1.7 Motions and collinearity	7
	1.8 A motion is affine linear on lines	7
	1.9 Motions are affine transformations	8
	1.10 Euclidean motions and orthogonal transformations	9
	1.11 Normal form of an orthogonal matrix	10
	1.11.1 The 2×2 rotation and reflection matrixes	10
	1.11.2 The general case	12
	1.12 Euclidean frames and motions	14
	1.13 Frames and motions of $\mathbb{E}^2$	14
	1.14 Every motion of $\mathbb{E}^2$ is a translation, rotation, reflection or glide	15
	1.15 Classification of motions of $\mathbb{E}^3$	17
	1.16 Sample theorems of Euclidean geometry	19
	1.16.1 Pons asinorum	19
	1.16.2 The angle sum of triangles	19
	1.16.3 Parallel lines and similar triangles	20
	1.16.4 Four centres of a triangle	21
	1.16.5 The Feuerbach 9-point circle	23
	Exercises	24
2	**Composing maps**	26
	2.1 Composition is the basic operation	26
	2.2 Composition of affine linear maps $\mathbf{x} \mapsto A\mathbf{x} + \mathbf{b}$	27

2.3 Composition of two reflections of $\mathbb{E}^2$ 27
2.4 Composition of maps is associative 28
2.5 Decomposing motions 28
2.6 Reflections generate all motions 29
2.7 An alternative proof of Theorem 1.14 31
2.8 Preview of transformation groups 31
Exercises 32

3   Spherical and hyperbolic non-Euclidean geometry 34
3.1   Basic definitions of spherical geometry 35
3.2   Spherical triangles and trig 37
3.3   The spherical triangle inequality 38
3.4   Spherical motions 38
3.5   Properties of S^2 like $\mathbb{E}^2$ 39
3.6   Properties of S^2 unlike $\mathbb{E}^2$ 40
3.7   Preview of hyperbolic geometry 41
3.8   Hyperbolic space 42
3.9   Hyperbolic distance 43
3.10 Hyperbolic triangles and trig 44
3.11 Hyperbolic motions 46
3.12 Incidence of two lines in $\mathcal{H}^2$ 47
3.13 The hyperbolic plane is non-Euclidean 49
3.14 Angular defect 51
        3.14.1 The first proof 51
        3.14.2 An explicit integral 51
        3.14.3 Proof by subdivision 53
        3.14.4 An alternative sketch proof 54
Exercises 56

4   Affine geometry 62
4.1 Motivation for affine space 62
4.2 Basic properties of affine space 63
4.3 The geometry of affine linear subspaces 65
4.4 Dimension of intersection 67
4.5 Affine transformations 68
4.6 Affine frames and affine transformations 68
4.7 The centroid 69
Exercises 69

5   Projective geometry 72
5.1   Motivation for projective geometry 72
        5.1.1   Inhomogeneous to homogeneous 72
        5.1.2   Perspective 73
        5.1.3   Asymptotes 73
        5.1.4   Compactification 75

5.2	Definition of projective space	75
5.3	Projective linear subspaces	76
5.4	Dimension of intersection	77
5.5	Projective linear transformations and projective frames of reference	77
5.6	Projective linear maps of $\mathbb{P}^1$ and the cross-ratio	79
5.7	Perspectivities	81
5.8	Affine space $\mathbb{A}^n$ as a subset of projective space $\mathbb{P}^n$	81
5.9	Desargues' theorem	82
5.10	Pappus' theorem	84
5.11	Principle of duality	85
5.12	Axiomatic projective geometry	86
	Exercises	88

6	**Geometry and group theory**	**92**
6.1	Transformations form a group	93
6.2	Transformation groups	94
6.3	Klein's Erlangen program	95
6.4	Conjugacy in transformation groups	96
6.5	Applications of conjugacy	98
	6.5.1 Normal forms	98
	6.5.2 Finding generators	100
	6.5.3 The algebraic structure of transformation groups	101
6.6	Discrete reflection groups	103
	Exercises	104

7	**Topology**	**107**
7.1	Definition of a topological space	108
7.2	Motivation from metric spaces	108
7.3	Continuous maps and homeomorphisms	111
	7.3.1 Definition of a continuous map	111
	7.3.2 Definition of a homeomorphism	111
	7.3.3 Homeomorphisms and the Erlangen program	112
	7.3.4 The homeomorphism problem	113
7.4	Topological properties	113
	7.4.1 Connected space	113
	7.4.2 Compact space	115
	7.4.3 Continuous image of a compact space is compact	116
	7.4.4 An application of topological properties	117
7.5	Subspace and quotient topology	117
7.6	Standard examples of glueing	118
7.7	Topology of $\mathbb{P}^n_{\mathbb{R}}$	121
7.8	Nonmetric quotient topologies	122
7.9	Basis for a topology	124

7.10 Product topology   126
7.11 The Hausdorff property   127
7.12 Compact versus closed   128
7.13 Closed maps   129
7.14 A criterion for homeomorphism   130
7.15 Loops and the winding number   130
    7.15.1 Paths, loops and families   131
    7.15.2 The winding number   133
    7.15.3 Winding number is constant in a family   135
    7.15.4 Applications of the winding number   136
    Exercises   137

8 **Quaternions, rotations and the geometry of
    transformation groups**   142
8.1 Topology on groups   143
8.2 Dimension counting   144
8.3 Compact and noncompact groups   146
8.4 Components   148
8.5 Quaternions, rotations and the geometry of SO(n)   149
    8.5.1 Quaternions   149
    8.5.2 Quaternions and rotations   151
    8.5.3 Spheres and special orthogonal groups   152
8.6 The group SU(2)   153
8.7 The electron spin in quantum mechanics   154
    8.7.1 The story of the electron spin   154
    8.7.2 Measuring spin: the Stern–Gerlach device   155
    8.7.3 The spin operator   156
    8.7.4 Rotate the device   157
    8.7.5 The solution   158
8.8 Preview of Lie groups   159
    Exercises   161

9 **Concluding remarks**   164
9.1 On the history of geometry   165
    9.1.1 Greek geometry and rigour   165
    9.1.2 The parallel postulate   165
    9.1.3 Coordinates versus axioms   168
9.2 Group theory   169
    9.2.1 Abstract groups versus transformation groups   169
    9.2.2 Homogeneous and principal homogeneous spaces   169
    9.2.3 The Erlangen program revisited   170
    9.2.4 Affine space as a torsor   171

9.3 Geometry in physics   172
    9.3.1   The Galilean group and Newtonian dynamics   172
    9.3.2   The Poincaré group and special relativity   173
    9.3.3   Wigner's classification: elementary particles   175
    9.3.4   The Standard Model and beyond   176
    9.3.5   Other connections   176
9.4 The famous trichotomy   177
    9.4.1   The curvature trichotomy in geometry   177
    9.4.2   On the shape and fate of the universe   178
    9.4.3   The snack bar at the end of the universe   179

Appendix A Metrics   180
Exercises   181

Appendix B Linear algebra   183
B.1 Bilinear form and quadratic form   183
B.2 Euclid and Lorentz   184
B.3 Complements and bases   185
B.4 Symmetries   186
B.5 Orthogonal and Lorentz matrixes   187
B.6 Hermitian forms and unitary matrixes   188
Exercises   189

References   190
Index   193

Figures

A coordinate model of space   *page* xiv

1.1   Triangle inequality   2
1.5   Angle with direction   6
1.6   Rigid body motion   6
1.9   Affine linear construction of $\lambda \mathbf{x} + \mu \mathbf{y}$ 9
1.11a A rotation in coordinates   11
1.11b The rotation and the reflection   11
1.13   The Euclidean frames P_0, P_1, P_2 and P_0', P_1', P_2' 14
1.14a Rot(O, θ) and Glide$(L, \mathbf{v})$ 15
1.14b Construction of glide   15
1.14c Construction of rotation   16
1.15a Twist $(L, \theta, \mathbf{v})$ and Rot-Refl (L, θ, Π) 17
1.15b A grid of parallel planes and their orthogonal lines   17
1.16a Pons asinorum   19
1.16b Sum of angles in a triangle is equal to π 20
1.16c Parallel lines fall on lines in the same ratio   20
1.16d Similar triangles   21
1.16e The centroid   21
1.16f The circumcentre   22
1.16g The orthocentre   22
1.16h The Feuerbach 9-point circle   23

2.3   Composite of two reflections   28
2.7   Composite of a rotation and a reflection   31

3.0   Plane-like geometry   35
3.2   Spherical trig   38
3.6   Overlapping segments of S^2 41
3.7   The hyperbola $t^2 = 1 + x^2$ and $t > 0$ 42
3.8   Hyperbolic space $\mathcal{H}^2$ 43

3.10   Hyperbolic trig   45
3.12   (a) Projection to the (x, y)-plane of the spherical lines $y = cz$
       (b) Projection to the (x, y)-plane of the hyperbolic lines $y = ct$ 48
3.13   The failure of the parallel postulate in $\mathcal{H}^2$ 50
3.14a The hyperbolic triangle $\triangle PQR$ with one ideal vertex   52
3.14b Area and angle sums are 'additive'   52
3.14c The subdivision of $\triangle PQR$.   54
3.14d The angular defect formula   55
3.14e Area is an additive function   56
3.14f Area is a monotonic function   56
3.15 $\mathcal{H}$-lines   60

4.2   Points, vectors and addition   64
4.3a   The affine construction of the line segment $[\mathbf{p}, \mathbf{q}]$ 66
4.3b   Parallel hyperplanes   66
4.7   The affine centroid   70
4.8   A weighted centroid   70

5.1a   A cube in perspective   74
5.1b   Perspective drawing   74
5.1c   Hyperbola and parabola   74
5.6a   The 3-transitive action of PGL(2) on $\mathbb{P}^1$ 80
5.6b   The cross-ratio $\{P, Q; R, S\}$ 80
5.8   The inclusion $\mathbb{A}^n \subset \mathbb{P}^n$ 82
5.9a   The Desargues configuration in $\mathbb{P}^2$ or $\mathbb{P}^3$ 83
5.9b   Lifting the Desargues configuration to $\mathbb{P}^3$ 84
5.10   The Pappus configuration   85
5.12a Axiomatic projective plane   87
5.12b Geometric construction of addition   88

6.0   The plan of Coventry market   93
6.4a   The conjugate rotation $g \operatorname{Rot}(P, \theta)g^{-1} = \operatorname{Rot}(g(P), g(\theta))$ 97
6.4b   Action of Aff(n) on vectors of $\mathbb{A}^n$ 98
6.6a   Kaleidoscope   104
6.6b   'Musée Grévin'   104

7.2a   Hausdorff property   110
7.2b $S^1 = [0, 1]$ with the ends identified   110
7.3a $(0, 1) \simeq \mathbb{R}$ 112
7.3b   Squaring the circle   112
7.4a   Path connected set   114
7.6a   The Möbius strip M 119
7.6b   The cylinder $S^1 \times [0, 1]$ 119

7.6c	The torus	120
7.6d	Surface with g handles	120
7.6e	Boundary and interior points	121
7.7	Topology of $\mathbb{P}^2_{\mathbb{R}}$: Möbius strip with a disc glued in	122
7.8a	The mousetrap topology	123
7.8b	Equivalence classes of quadratic forms $ax^2 + 2bxy + cy^2$	124
7.10	Balls for product metrics	126
7.12	Separating a point from a compact subset	128
7.13a	Closed map	129
7.13b	Nonclosed map	129
7.15a	Continuous family of paths	131
7.15b	D^* covered by overlapping open radial sectors	134
7.15c	Overlapping intervals	134
7.16a	Glueing patterns on the square	140
7.16b	The surface with two handles and the 12-gon	140
8.0	The geometry of the group of planar rotations	143
8.7a	The Stern–Gerlach experiment	155
8.7b	The modified Stern–Gerlach device	156
8.7c	Two identical SG devices	156
8.7d	Two different SG devices	157
9.1a	The parallel postulate. To meet or not to meet?	166
9.1b	The parallel postulate in the Euclidean plane	166
9.1c	The 'parallel postulate' in spherical geometry	168
9.4a	The cap, flat plane and Pringle's chip	178
9.4b	The genus trichotomy $g = 0$, $g = 1$, $g \geq 2$ for oriented surfaces	178
A.1	The bear	182

Preface

What is geometry about?

Geometry 'measuring the world' attempts to describe and understand space around us and all that is in it. It is the central activity and main driving force in many branches of math and physics, and offers a whole range of views on the nature and meaning of the universe. This book treats geometry in a wide context, including a wealth of relations with surrounding areas of math and other aspects of human experience.

Any discussion of geometry involves tension between the twin ideals of intuition and precision. *Descriptive* or *synthetic* geometry takes as its starting point our ideas and experience of the observed world, and treats geometric objects such as lines and shapes as objects in their own right. For example, a line could be the path of a light ray in space; you can envisage comparing line segments or angles by 'moving' one over another, thus giving rise to notions of 'congruent' figures, equal lengths, or equal angles that are independent of any quantitative measurement. If A, B, C are points along a line segment, what it means for B to be *between* A and C is an idea hard-wired into our consciousness. While descriptive geometry is intuitive and natural, and can be made mathematically rigorous (and, of course, Euclidean geometry was studied in these terms for more than two millennia, compare 9.1), this is not my main approach in this book.

My treatment centres rather on *coordinate* geometry. This uses Descartes' idea (1637) of measuring distances to view points of space and geometric quantities in terms of numbers, with respect to a fixed origin, using intuitive ideas such as 'a bit to the right' or 'a long way up' and using them quantitatively in a systematic and precise way. In other words, I set up the (x, y)-plane $\mathbb{R}^2$, the (x, y, z)-space $\mathbb{R}^3$ or whatever I need, and use it as a mathematical model of the plane (space, etc.), for the purposes of calculations. For example, to plan the layout of a car park, I might map it onto a sheet of paper or a computer screen, pretending that pairs (x, y) of real numbers correspond to points of the surface of the earth, at least in the limited region for which I have planning permission. Geometric constructions, such as drawing an even rectangular grid or planning the position of the ticket machines to ensure the maximum aggravation to customers, are easier to make in the model than in real

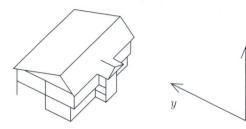

A coordinate model of space.

life. We admit possible drawbacks of our model, but its use divides any problem into calculations within the model, and considerations of how well it reflects the practical world.

Topology is the youngster of the geometry family. Compared to its venerable predecessors, it really only got going in the twentieth century. It dispenses with practically all the familiar quantities central to other branches of geometry, such as distance, angles, cross-ratios, and so on. If you are tempted to the conclusion that there is not much left for topology to study, think again. Whether two loops of string are linked or not does not depend on length or shape or perspective; if that seems too simple to be a serious object of study, what about the linking or knotting of strands of DNA, or planning the over- and undercrossings on a microchip? The higher dimensional analogues of disconnecting or knotting are highly nontrivial and not at all intuitive to denizens of the lower dimensions such as ourselves, and cannot be discussed without formal apparatus. My treatment of topology runs briefly through abstract *point-set topology*, a fairly harmless generalisation of the notion of continuity from a first course on analysis and metric spaces. However, my main interest is in topology as *rubber-sheet geometry*, dealing with manifestly geometric ideas such as closed curves, spheres, the torus, the Möbius strip and the Klein bottle.

Change of coordinates, motions, group theory and the Erlangen program

Descartes' idea to use numbers to describe points in space involves the choice of a *coordinate system* or *coordinate frame*: an origin, together with axes and units of length along the axes. A recurring theme of all the different geometries in this book is the question of what a coordinate frame is, and what I can get out of it. While coordinates provide a convenient framework to discuss points, lines, and so on, it is a basic requirement that any meaningful statement in geometry is *independent of the choice of coordinates*. That is, coordinate frames are a humble technical aid in determining the truth, and are not allowed the dignity of having their own meaning.

Changing from one coordinate frame to another can be viewed as a *transformation* or *motion*: I can use a motion of space to align the origin and coordinate axes of two coordinate systems. A statement that remains true under any such motion is independent of the choice of coordinates. Felix Klein's 1872 *Erlangen program* formalises this relation between geometric properties and changes of coordinates by defining geometry to be the study of properties invariant under allowed coordinate changes, that is, invariant under a *group of transformations*. This approach is closely related to the point of view of special relativity in theoretical physics (Einstein, 1905), which insists that the laws of physics must be invariant under Lorentz transformations.

This course discusses several different geometries: in some case the spaces themselves are different (for example, the sphere and the plane), but in others the difference is purely in the conventions I make about coordinate changes. Metric geometries such as Euclidean and hyperbolic non-Euclidean geometry include the notions of distance between two points and angle between two lines. The allowed transformations are rigid motions (isometries or congruences) of Euclidean or hyperbolic space. Affine and projective geometries consider properties such as collinearity of points, and the typical group is the general linear group $GL(n)$, the group of invertible $n \times n$ matrixes. Projective geometry presents an interesting paradox: while its mathematical treatment involves what may seem to be quite arcane calculations, your brain has a sight driver that carries out projective transformations by the thousand every time you recognise an object in perspective, and does so unconsciously and practically instantaneously.

The sets of transformations that appear in topology, for example the set of all continuous one-to-one maps of the interval [0, 1] to itself, or the same thing for the circle S^1 or the sphere S^2, are of course too big for us to study by analogy with transformation groups such as $GL(n)$ or the Euclidean group, whose elements depend on finitely many parameters. In the spirit of the Erlangen program, properties of spaces that remain invariant under such a huge set of equivalences must be correspondingly coarse. I treat a few basic topological properties such as compactness, connectedness, winding number and simple connectedness that appear in many different areas of analysis and geometry. I use these simple ideas to motivate the central problem of topology: how to distinguish between topologically different spaces? At a more advanced level, topology has developed systematic invariants that apply to this problem, notably the fundamental group and homology groups. These are invariants of spaces that are the same for topologically equivalent spaces. Thus if you can calculate one of these invariants for two spaces (for example, a disc and a punctured disc) and prove that the answers are different, then the spaces are certainly not topologically equivalent. You may want to take subsequent courses in topology to become a real expert, and this course should serve as a useful guide in this.

Geometry in applications

Although this book is primarily intended for use in a math course, and the topics are oriented towards the theoretical foundations of geometry, I must stress that the math ideas discussed here are applicable in different ways, basic or sophisticated, as stated or with extra development, on their own or in combination with other disciplines, Euclidean or non-Euclidean, metric or topological, to a huge variety of scientific and technological problems in the modern world. I discuss in Chapter 8 the quantum mechanical description of the electron that illustrates a fundamental application of the ideas of group theory and topology to the physics of elementary particles. To move away from basic to more applied science, let me mention a few examples from technology. The typesetting and page layout software now used throughout the newspaper and publishing industry, as well as in the computer rooms of most university departments, can obviously not exist without a knowledge of basic coordinate geometry: even a primary instruction such as 'place letter A or box B, scaled by such-and-such a factor, slanted at such-and-such an angle, at such-and-such a point on the page' involves affine transformations. Within the same industry, computer typefaces themselves are designed using Bezier curves. The geometry used in robotics is more sophisticated. The technological aim is, say, to get a robot arm holding a spanner into the right position and orientation, by adjusting some parameters, say, angles at joints or lengths of rods. This translates in a fairly obvious way into the geometric problem of parametrising a piece of the Euclidean group; but the solution or approximate solution of this problem is hard, involving the topology and analysis of manifolds, algebraic geometry and singularity theory. The computer processing of camera images, whose applications include missile guidance systems, depends among other things on projective transformations (I say this for the benefit of students looking for a career truly worthy of their talents and education). Although scarcely having the same nobility of purpose, similar techniques apply in ultrasonic scanning used in antenatal clinics; here the geometric problem is to map the variations in density in a 3-dimensional medium onto a 2-dimensional computer screen using ultrasonic radar, from which the human eye can easily make out salient features. By a curious coincidence, 3 hours before I, the senior author, gave the first lecture of this course in January 1989, I was at the maternity clinic of Walsgrave hospital Coventry looking at just such an image of a 16-week old foetus, now my third daughter Murasaki.

About this book

Who the book is for

This book is intended for the early years of study of an undergraduate math course. For the most part, it is based on a second year module taught at Warwick over many years, a module that is also taken by first and third year math students, and by students from the math/physics course. You will find the book accessible if you are familiar with most of the following, which is standard material in first and second year math courses.

Coordinate geometry   How to express lines and circles in $\mathbb{R}^2$ in terms of coordinates, and calculate their points of intersection; some idea of how to do the same in $\mathbb{R}^3$ and maybe $\mathbb{R}^n$ may also be helpful.

Linear algebra   Vector spaces and linear maps over $\mathbb{R}$ and $\mathbb{C}$, bases and matrixes, change of bases, eigenvalues and eigenvectors. This is the only major piece of math that I take for granted. The examples and exercises make occasional reference to vector spaces over fields other than $\mathbb{R}$ or $\mathbb{C}$ (such as finite fields), but you can always omit these bits if they make you uncomfortable.

Multilinear algebra   Bilinear and quadratic forms, and how to express them in matrix terms; also Hermitian forms. I summarise all the necessary background material in Appendix B.

Metric spaces   Some prior familiarity with the first ideas of a metric space course would not do any harm, but this is elementary material, and Appendix A contains all that you need to know.

Group theory   I have gone to some trouble to develop from first principles all the group theory that I need, with the intention that my book can serve as a first introduction to transformation groups and the notions of abstract group theory if you have never seen these. However, if you already have some idea of basic things such as composition laws, subgroups, cosets and the symmetric group, these will come in handy as motivation. If you prefer to see a conventional introduction to group theory, there are any number of textbooks, for example Green [10] or Ledermann [14]. If you intend to study group theory beyond the introductory stage, I strongly recommend Artin [1] or Segal [22]. My ideological slant on this issue is discussed in more detail in 9.2.

How to use the book   Although the thousands queueing impatiently at supermarkets and airport bookshops to get their hands on a copy of this book for vacation reading was strong motivation for me in writing it, experience suggests the harsher view of reality: at least some of my readers may benefit from coercion in the form of an organised lecture course.

Experience from teaching at Warwick shows that Chapters 1–6 make a reasonably paced 30 hour second year lecture course. Some more meat could be added to subjects that the lecturer or students find interesting; reflection groups following Coxeter [5], Chapter 4 would be one good candidate. Topics from Chapters 7–8 or the further topics of Chapter 9 could then profitably be assigned to students as essay or project material. An alternative course oriented towards group theory could start with affine and Euclidean geometry and some elements of topology (maybe as a refresher), and concentrate on Chapters 3, 6 and 8, possibly concluding with some material from Segal [22]. This would provide motivation and techniques to study matrix groups from a geometric point of view, one often ignored in current texts.

The author's identity crisis   I want the book to be as informal as possible in style. To this end, I always refer to the student as 'you', which has the additional advantage that it is independent of your gender and number. I also refer to myself by the first person singular, despite the fact that there are two of me. Each of me has lectured the material many times, and is used to taking personal responsibility for the truth of my assertions. My model is van der Waerden's style, who always wrote the crisp 'Ich behaupte...' (often when describing results he learned from Emmy Noether or Emil Artin's lectures). I leave you to imagine the speaker as your ideal teacher, be it a bearded patriarch or a fresh-faced bespectacled Central European intellectual.

Acknowledgements   A second year course with the title 'Geometry' or 'Geometry and topology' has been given at Warwick since the 1960s. It goes without saying that my choice of material, and sometimes the material itself, is taken in part from the experience of colleagues, including John Jones, Colin Rourke, Brian Sanderson; David Epstein has also provided some valuable material, notably in the chapter on hyperbolic geometry. I have also copied material consciously or unconsciously from several of the textbooks recommended for the course, in particular Coxeter [5], Rees [19], Nikulin and Shafarevich [18] and Feynman [7]. I owe special thanks to Katrin Wendland, the most recent lecturer of the Warwick course MA243 Geometry, who has provided a detailed criticism of my text, thereby saving me from a variety of embarrassments.

Disclaimer   Wen solche Lehren nicht erfreun,
Verdienet nicht ein Mensch zu sein.

From Sarastro's aria, The Magic Flute, II.3.

This is an optional course. If you don't like my teaching, please deregister before the deadline.

1 Euclidean geometry

This chapter discusses the geometry of n-dimensional Euclidean space $\mathbb{E}^n$, together with its distance function. The distance gives rise to other notions such as angles and congruent triangles. Choosing a Euclidean coordinate frame, consisting of an origin O and an orthonormal basis of vectors out of O, leads to a description of $\mathbb{E}^n$ by coordinates, that is, to an identification $\mathbb{E}^n = \mathbb{R}^n$.

A map of Euclidean space preserving Euclidean distance is called a *motion* or *rigid body motion*. Motions are fun to study in their own right. My aims are

(1) to describe motions in terms of linear algebra and matrixes;
(2) to find out how many motions there are;
(3) to describe (or classify) each motion individually.

I do this rather completely for $n = 2, 3$ and some of it for all n. For example, the answer to (2) is that all points of $\mathbb{E}^n$, and all sets of orthonormal coordinate frames at a point, are equivalent: given any two frames, there is a unique motion taking one to the other. In other words, any point can serve as the origin, and any set of orthogonal axes as the coordinate frames. This is the geometric and philosophical principle that space is homogeneous and isotropic (the same viewed from every point and in every direction). The answer to (3) in $\mathbb{E}^2$ is that there are four types of motions: translations and rotations, reflections and glides (Theorem 1.14).

The chapter concludes with some elementary sample theorems of plane Euclidean geometry.

1.1 The metric on $\mathbb{R}^n$

Throughout the book, I write $\mathbb{R}^n$ for the vector space of n-tuples $(x_1, \ldots, x_n)$ of real numbers. I start by discussing its metric geometry. The familiar Euclidean distance function on $\mathbb{R}^n$ is defined by

$$|\mathbf{x} - \mathbf{y}| = \sqrt{\left(\sum (x_i - y_i)^2\right)}, \quad \text{where } \mathbf{x} = \begin{pmatrix} x_1 \\ \vdots \\ x_n \end{pmatrix} \text{ and } \mathbf{y} = \begin{pmatrix} y_1 \\ \vdots \\ y_n \end{pmatrix}. \tag{1}$$

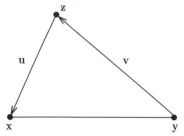

Figure 1.1   Triangle inequality.

The relationship between this distance function and the Euclidean inner product (or dot product) $\mathbf{x} \cdot \mathbf{y} = \sum x_i y_i$ on $\mathbb{R}^n$ is discussed in Appendix B.2. The more important point is that the Euclidean distance (1) is a metric on $\mathbb{R}^n$. If you have not yet met the idea of a *metric* on a set X, see Appendix A; for now recall that it is a distance function $d(\mathbf{x}, \mathbf{y})$ satisfying positivity, symmetry and the triangle inequality. Both the positivity $|\mathbf{x} - \mathbf{y}| \geq 0$ and symmetry $|\mathbf{x} - \mathbf{y}| = |\mathbf{y} - \mathbf{x}|$ are immediate, so the point is to prove the triangle inequality (Figure 1.1).

Theorem (Triangle inequality)

$$|\mathbf{x} - \mathbf{y}| \leq |\mathbf{x} - \mathbf{z}| + |\mathbf{z} - \mathbf{y}|, \quad \textit{for all } \mathbf{x}, \mathbf{y}, \mathbf{z} \in \mathbb{R}^n, \tag{2}$$

with equality if and only if $\mathbf{z} = \mathbf{x} + \lambda(\mathbf{y} - \mathbf{x})$ *for* λ *a real number between* 0 *and* 1.

Proof   Set $\mathbf{x} - \mathbf{z} = \mathbf{u}$ and $\mathbf{z} - \mathbf{y} = \mathbf{v}$ so that $\mathbf{x} - \mathbf{y} = \mathbf{u} + \mathbf{v}$; then (2) is equivalent to

$$\sqrt{\sum u_i^2} + \sqrt{\sum v_i^2} \geq \sqrt{\left(\sum (u_i + v_i)^2 \right)}. \tag{3}$$

Note that both sides are nonnegative, so that squaring, one sees that (3) is equivalent to

$$\sum u_i^2 + \sum v_i^2 + 2\sqrt{\left(\sum u_i^2 \cdot \sum v_i^2 \right)} \geq \sum (u_i + v_i)^2$$
$$= \sum u_i^2 + \sum v_i^2 + 2 \sum u_i v_i. \tag{4}$$

Cancelling terms, one sees that (4) is equivalent to

$$\sqrt{\left(\sum u_i^2 \cdot \sum v_i^2 \right)} \geq \sum u_i v_i. \tag{5}$$

If the right-hand side is negative then (5), hence also (2), is true and strict. If the right-hand side of (5) is ≥ 0 then it is again permissible to square both sides, giving

$$\sum u_i^2 \cdot \sum v_j^2 \geq \left(\sum u_i v_i \right) \left(\sum u_j v_j \right). \tag{6}$$

You will see at once what is going on if you write this out explicitly for $n = 2$ and expand both sides. For general n, the trick is to use two different dummy indexes i, j as in (6): expanding and cancelling gives that (6) is equivalent to

$$\sum_{i > j} (u_i v_j - u_j v_i)^2 \geq 0. \tag{7}$$

Now (7) is true, so retracing our steps back through the argument gives that (2) is true. Finally, equality in (2) holds if and only if $u_i v_j = u_j v_i$ for all i, j (from (7)) and $\sum u_i v_i \geq 0$ (from the right-hand side of (5)); that is, $\mathbf{u}$ and $\mathbf{v}$ are proportional, $\mathbf{u} = \mu \mathbf{v}$ with $\mu \geq 0$. Rewriting this in terms of $\mathbf{x}$, $\mathbf{y}$, $\mathbf{z}$ gives the conclusion.   QED

1.2 Lines and collinearity in $\mathbb{R}^n$

There are several ways of defining a line (already in the usual x, y plane $\mathbb{R}^2$); I choose one definition for $\mathbb{R}^n$.

Definition   Let $\mathbf{u} \in \mathbb{R}^n$ be a fixed point and $\mathbf{v} \in \mathbb{R}^n$ a nonzero direction vector. The *line L* starting at $\mathbf{u} \in \mathbb{R}^n$ with direction vector $\mathbf{v}$ is the set

$$L := \{\mathbf{u} + \lambda \mathbf{v} \mid \lambda \in \mathbb{R}\} \subset \mathbb{R}^n.$$

Three distinct points $\mathbf{x}$, $\mathbf{y}$, $\mathbf{z} \in \mathbb{R}^n$ are *collinear* if they are on a line.

If I choose the starting point $\mathbf{x}$, and the direction vector $\mathbf{v} = \mathbf{y} - \mathbf{x}$, then $L = \{(1 - \lambda)\mathbf{x} + \lambda\mathbf{y}\}$. To say that distinct points $\mathbf{x}$, $\mathbf{y}$, $\mathbf{z}$ are collinear means that $\mathbf{z} = \{(1 - \lambda)\mathbf{x} + \lambda\mathbf{y}\}$ for some λ. Writing

$$[\mathbf{x}, \mathbf{y}] = \{\mathbf{x} + \lambda(\mathbf{y} - \mathbf{x}) \mid 0 \leq \lambda \leq 1\}$$

for the line segment between $\mathbf{x}$ and $\mathbf{y}$, the possible orderings of $\mathbf{x}$, $\mathbf{y}$, $\mathbf{z}$ on the line L are controlled by

$$\left. \begin{array}{c} \lambda \leq 0 \\ 0 \leq \lambda \leq 1 \\ 1 \leq \lambda \end{array} \right\} \iff \left\{ \begin{array}{c} \mathbf{x} \in [\mathbf{z}, \mathbf{y}] \\ \mathbf{z} \in [\mathbf{x}, \mathbf{y}] \\ \mathbf{y} \in [\mathbf{x}, \mathbf{z}]. \end{array} \right.$$

Together with the triangle inequality Theorem 1.1, this proves the following result.

Corollary   *Three distinct points* $\mathbf{x}$, $\mathbf{y}$, $\mathbf{z} \in \mathbb{R}^n$ *are collinear if and only if (after a permutation of* $\mathbf{x}$, $\mathbf{y}$, $\mathbf{z}$ *if necessary)*

$$|\mathbf{x} - \mathbf{y}| + |\mathbf{y} - \mathbf{z}| = |\mathbf{x} - \mathbf{z}|.$$

In other words, collinearity is determined by the metric.

1.3   Euclidean space $\mathbb{E}^n$

After these preparations, I am ready to introduce the main object of study: *Euclidean n-space* $(\mathbb{E}^n, d)$ is a metric space (with metric d) for which there exists a bijective map $\mathbb{E}^n \to \mathbb{R}^n$, such that if $P, Q \in \mathbb{E}^n$ are mapped to $\mathbf{x}, \mathbf{y} \in \mathbb{R}^n$ then

$$d(P, Q) = |\mathbf{y} - \mathbf{x}|.$$

In other words, $(\mathbb{E}^n, d)$ is isometric to the vector space $\mathbb{R}^n$ with its usual distance function, if you like this kind of language.

Since lines and collinearity in $\mathbb{R}^n$ are characterised purely in terms of the Euclidean distance function, these notions carry over to $\mathbb{E}^n$ without any change: three points of $\mathbb{E}^n$ are collinear if they are collinear for some isometry $\mathbb{E}^n \to \mathbb{R}^n$ (hence for all possible isometries); the lines of $\mathbb{E}^n$ are the lines of $\mathbb{R}^n$ under any such identification. For example, for points $P, Q \in \mathbb{E}^n$, the line segment $[P, Q] \subset \mathbb{E}^n$ is the set

$$[P, Q] = \left\{ R \in \mathbb{E}^n \;\middle|\; d(P, R) + d(R, Q) = d(P, Q) \right\} \subset \mathbb{E}^n.$$

Remark   The main point of the definition of $\mathbb{E}^n$ is that the map $\mathbb{E}^n \to \mathbb{R}^n$ identifying the metrics is not fixed throughout the discussion; I only insist that one such isometry should exist. A particular choice of identification preserving the metric is referred to as a choice of *(Euclidean) coordinates*. Points of $\mathbb{E}^n$ will always be denoted by capital letters P, Q; once I choose a bijection, the points acquire *coordinates* $P = (x_1, \ldots, x_n)$. In particular, any coordinate system distinguishes one point of $\mathbb{E}^n$ as the origin $(0, \ldots, 0)$; however, different identifications pick out different points of $\mathbb{E}^n$ as their origin. If you want to have a Grand Mosque of Mecca or a Greenwich Observatory, you must either receive it by Divine Grace or make a deliberate extra choice. The idea of space ought to make sense without a coordinate system, but you can always fix one if you like.

You can also look at this process from the opposite point of view. Going from $\mathbb{R}^n$ to $\mathbb{E}^n$, I forget the distinguished origin $0 \in \mathbb{R}^n$, the standard coordinate system, and the vector space structure of $\mathbb{R}^n$, remembering only the distance and properties that can be derived from it.

1.4   Digression: shortest distance

As just shown, the metric of Euclidean space $\mathbb{E}^n$ determines the lines. This section digresses to discuss the idea summarised in the well known cliché 'a straight line is the shortest distance between two points'; while logically not absolutely essential in this chapter, this idea is important in the philosophy of Euclidean geometry (as well as spherical and hyperbolic geometry).

Principle   *The distance $d(P, Q)$ between two points $P, Q \in \mathbb{E}^n$ is the length of the shortest curve joining P and Q. The line segment $[P, Q]$ is the unique shortest curve joining P, Q.*

Sketch proof   This looks obvious: if a curve C strays off the straight and narrow to some point $R \notin [P, Q]$, its length is at least

$$d(P, R) + d(R, Q) > d(P, Q).$$

The statement is, however, more subtle: for instance, it clearly does not make sense without a definition of a curve C and its length. A curve C in $\mathbb{E}^n$ from P to Q is a family of points $R_t \in \mathbb{E}^n$, depending on a 'time variable' t such that $R_0 = P$ and $R_1 = Q$. Clearly R_t should at least be a *continuous* function of t – if you allow instantaneous 'teleporting' between far away points, you can obviously get arbitrarily short paths.

The proper definition of curves and lengths of curves belongs to differential geometry or analysis. Given a 'sufficiently smooth' curve, you can define its length as the integral $\int_C \mathrm{d}s$ along C of the infinitesimal arc length $\mathrm{d}s$, given by $\mathrm{d}s^2 = \sum_{i=1}^{n} \mathrm{d}x_i^2$. Alternatively, you can mark out successive points $P = R_0, R_1, \ldots, R_{N+1} = Q$ along the curve, view the sum $\sum_{i=0}^{N} d(R_i, R_{i+1})$ as an approximation to the length of C, and define the length of C to be the supremum taken over all such piecewise linear approximations. To avoid the analytic details (which are not at all trivial!), I argue under the following weak assumption: under any reasonable definition of the length of C,

for any $\varepsilon > 0$, the curve C can be closely approximated by a piecewise linear path made up of short intervals $[P, R_1]$, $[R_1, R_2]$, etc., such that

length of $C \geq$ sum of the lengths of the intervals $- \varepsilon$.

However, by the triangle inequality $d(P, R_2) \leq d(P, R_1) + d(R_1, R_2)$, so that the piecewise linear path can only get shorter if I omit R_1. Dealing likewise with R_2, R_3, etc., it follows that the length of C is $\geq d(P, Q) - \varepsilon$. Since this is true for any $\varepsilon > 0$, it follows that the length of C is $\geq d(P, Q)$. Thus the line interval $[P, Q]$ joining P, Q is the shortest path between them, and its length is $d(P, Q)$ by definition.   QED

1.5 Angles

The geometric significance of the Euclidean inner product $\mathbf{x} \cdot \mathbf{y} = \sum_{i=1}^{n} x_i y_i$ on $\mathbb{R}^n$ (Section B.2) is that the inner product measures the size of the *angle* $\angle\mathbf{xyz}$ based at $\mathbf{y}$ for $\mathbf{x}, \mathbf{y}, \mathbf{z} \in \mathbb{R}^n$:

$$\cos(\angle\mathbf{xyz}) = \frac{(\mathbf{x} - \mathbf{y}) \cdot (\mathbf{z} - \mathbf{y})}{|\mathbf{x} - \mathbf{y}||\mathbf{z} - \mathbf{y}|}. \tag{8}$$

By convention, I usually choose the angle to be between 0 and π. In particular, the vectors $\mathbf{x} - \mathbf{y}$, $\mathbf{z} - \mathbf{y}$ are *orthogonal* if $(\mathbf{x} - \mathbf{y}) \cdot (\mathbf{z} - \mathbf{y}) = 0$.

The notion of angle is easily transported to Euclidean space $\mathbb{E}^n$. Namely, the angle spanned by three points of $\mathbb{E}^n$ is defined to be the corresponding angle in $\mathbb{R}^n$ under a choice of coordinates. The angle is independent of this choice, because the inner product in $\mathbb{R}^n$ is determined by the quadratic form (Proposition B.1), and so ultimately

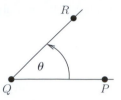

Figure 1.5   Angle with direction.

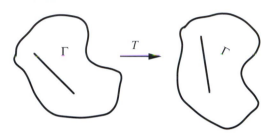

Figure 1.6   Rigid body motion.

by the metric of $\mathbb{E}^n$. In other words, the notion of angle is intrinsic to the geometry of $\mathbb{E}^n$.

There is one final issue to discuss regarding angles that is specific to the Euclidean plane $\mathbb{E}^2$. Namely, *once I fix a specific coordinate system* in $\mathbb{E}^2$, angles $\angle PQR$ acquire a *direction* as well as a size, once we agree (as we usually do) that an anticlockwise angle counts as positive, and a clockwise angle as negative. In Figure 1.5,

$$\angle PQR = -\angle RQP = \theta.$$

Under this convention, angles lie between $-\pi$ and π. Of course formula (8) does not reveal the sign as $\cos\theta = \cos(-\theta)$. It is important to realise that the direction of the angle is not intrinsic to $\mathbb{E}^2$, since a different choice of coordinates may reverse the sign.

1.6 Motions

A *motion* $T : \mathbb{E}^n \to \mathbb{E}^n$ is a transformation that preserves distances; that is, T is bijective, and

$$d(T(P), T(Q)) = d(P, Q) \quad \text{for all } P, Q \in \mathbb{E}^n.$$

The word motion is short for *rigid body motion*; it is alternatively called *isometry* or *congruence*. To say that T preserves distances means that there is 'no squashing or bending', hence the term rigid body motion; see Figure 1.6.

I study motions in terms of coordinates. After a choice of coordinates $\mathbb{E}^n \to \mathbb{R}^n$, a motion T gives rise to a map $T : \mathbb{R}^n \to \mathbb{R}^n$, its coordinate expression, which satisfies

$$|T(\mathbf{x}) - T(\mathbf{y})| = |\mathbf{x} - \mathbf{y}| \quad \text{for all } \mathbf{x}, \mathbf{y} \in \mathbb{R}^n.$$

The first thing I set out to do is to get from the abstract 'preserves distance' definition of a motion to the concrete coordinate expression $T(\mathbf{x}) = A\mathbf{x} + \mathbf{b}$ with A an orthogonal matrix. In the case of the Euclidean plane $\mathbb{E}^2$, the result is even more concrete; A is either a rotation matrix or a reflection matrix:

$$\begin{pmatrix} \cos\theta & -\sin\theta \\ \sin\theta & \cos\theta \end{pmatrix} \quad \text{or} \quad \begin{pmatrix} \cos\theta & \sin\theta \\ \sin\theta & -\cos\theta \end{pmatrix}.$$

1.7 Motions and collinearity

Proposition *A motion $T: \mathbb{E}^n \to \mathbb{E}^n$ preserves collinearity of points, so it takes lines to lines.*

Proof $P, Q, R \in E^n$ are collinear if and only if, possibly after a permutation of P, Q, R,

$$d(P, R) + d(R, Q) = d(P, Q).$$

But T preserves the distance function, so this happens if and only if, possibly after a permutation,

$$d(T(P), T(R)) + d(T(R), T(Q)) = d(T(P), T(Q))$$

which is equivalent to $T(P), T(Q), T(R)$ collinear. QED

The point is of course that, as we saw in 1.3, collinearity can be defined purely in terms of distance; since a motion T preserves distance, it preserves collinearity.

1.8 A motion is affine linear on lines

Proposition *If $T: \mathbb{R}^n \to \mathbb{R}^n$ is a motion expressed in coordinates, then*

$$T((1-\lambda)\mathbf{x} + \lambda\mathbf{y}) = (1-\lambda)T(\mathbf{x}) + \lambda T(\mathbf{y})$$

for all $\mathbf{x}, \mathbf{y} \in \mathbb{R}^n$ and all $\lambda \in \mathbb{R}$.

Proof A calculation based on the same idea as the previous proof: let $\mathbf{z} = (1-\lambda)\mathbf{x} + \lambda\mathbf{y}$. If $\mathbf{x} = \mathbf{y}$ there is nothing to prove; set $d = |\mathbf{x} - \mathbf{y}|$. Assume first that $\lambda \in [0, 1]$, so that $\mathbf{z} \in [\mathbf{x}, \mathbf{y}]$. Then, as in the previous proposition, $T(\mathbf{z}) \in [T(\mathbf{x}), T(\mathbf{y})]$, so $T(\mathbf{z}) = (1-\mu)T(\mathbf{x}) + \mu T(\mathbf{y})$ for some μ. But $|\mathbf{z} - \mathbf{x}| = \lambda d$, so $T(\mathbf{z})$ is the point at distance $(1-\lambda)d$ from $T(\mathbf{y})$ and λd from $T(\mathbf{x})$, that is, $\mu = \lambda$.

If $\lambda < 0$, say, then $\mathbf{x} \in [\mathbf{y}, \mathbf{z}]$ with $\mathbf{x} = (1-\lambda')\mathbf{y} + \lambda'\mathbf{z}$ and the same argument gives $T(\mathbf{x}) = (1-\lambda')T(\mathbf{y}) + \lambda'T(\mathbf{z})$, and you can derive the statement as an easy exercise. (The point is to write λ' as a function of λ; see Exercise 1.3.) QED

1.9   Motions are affine transformations

Definition   A map $T : \mathbb{E}^n \to \mathbb{E}^n$ is an *affine transformation* if it is given in a co-ordinate system by $T(\mathbf{x}) = A\mathbf{x} + \mathbf{b}$, where $A = (a_{ij})$ is an $n \times n$ matrix with nonzero determinant and $\mathbf{b} = (b_i)$ a vector; in more detail,

$$\mathbf{x} = (x_i) \mapsto \mathbf{y} = \left(\sum_{j=1}^{n} a_{ij} x_j + b_i \right), \quad \text{or} \quad \begin{pmatrix} x_1 \\ \vdots \\ x_n \end{pmatrix} \mapsto A \begin{pmatrix} x_1 \\ \vdots \\ x_n \end{pmatrix} + \begin{pmatrix} b_1 \\ \vdots \\ b_n \end{pmatrix}. \quad (9)$$

Proposition   *Let $T : \mathbb{E}^n \to \mathbb{E}^n$ be any map. Equivalent conditions:*

(1)   *T is given in some coordinate system by $T(\mathbf{x}) = A\mathbf{x} + \mathbf{b}$ for A an $n \times n$ matrix.*
(2)   *For all vectors $\mathbf{x}, \mathbf{y} \in \mathbb{R}^n$ and all $\lambda, \mu \in \mathbb{R}$ we have*

$$T\left(\lambda\mathbf{x} + \mu\mathbf{y}\right) - T(0) = \lambda\left(T(\mathbf{x}) - T(0)\right) + \mu\left(T(\mathbf{y}) - T(0)\right).$$

(3)   *For all $\mathbf{x}, \mathbf{y} \in \mathbb{R}^n$ and all $\lambda \in \mathbb{R}$*

$$T\left((1 - \lambda)\mathbf{x} + \lambda\mathbf{y}\right) = (1 - \lambda)T(\mathbf{x}) + \lambda T(\mathbf{y}).$$

that is, T is affine linear when restricted to any line.

Discussion   The point of the proposition is that condition (3) is a priori much weaker than the other two; it only requires that the map T is affine when restricted to lines. Note also that using the origin 0 in (2) seems to go against my expressed wisdom that there is no distinguished origin in the geometry of $\mathbb{E}^n$. However, recall that any point $P \in \mathbb{E}^n$ can serve as origin after a suitable translation.

Proof   (1) $\implies$ (2) is an easy exercise. (2) means exactly that if after performing T we translate by minus the vector $\mathbf{b} = T(0)$ to take $T(0)$ back to 0, then T becomes a linear map of vector spaces. Thus (2) $\implies$ (1) comes from the standard result of linear algebra expressing a linear map as a matrix.

(3) is just the particular case $\lambda + \mu = 1$ of (2). Thus the point of the proposition is to prove (3) $\implies$ (2).

Statement (2) concerns only the 2-dimensional vector subspace spanned by $\mathbf{x}, \mathbf{y} \in V$. We use statement (3) on the two lines $0\mathbf{x}$ and $0\mathbf{y}$ (see Figure 1.9), to get

$$T(2\lambda\mathbf{x}) = (1 - 2\lambda)T(0) + 2\lambda T(\mathbf{x})$$

and

$$T(2\mu\mathbf{y}) = (1 - 2\mu)T(0) + 2\mu T(\mathbf{y}).$$

Now apply (3) again to the line spanned by $2\lambda\mathbf{x}$ and $2\mu\mathbf{y}$:

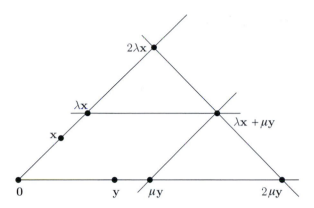

Figure 1.9 Affine linear construction of $\lambda\mathbf{x} + \mu\mathbf{y}$.

$$T\left(\lambda\mathbf{x} + \mu\mathbf{y}\right) = \frac{1}{2}T(2\lambda\mathbf{x}) + \frac{1}{2}T(2\mu\mathbf{y})$$

$$= \frac{1}{2}\left((1 - 2\lambda)T(0) + 2\lambda T(\mathbf{x})\right) + \frac{1}{2}\left((1 - 2\mu)T(0) + 2\mu T(\mathbf{y})\right)$$

$$= T(0) + \lambda\left(T(\mathbf{x}) - T(0)\right) + \mu\left(T(\mathbf{y}) - T(0)\right),$$

as required.   QED

Remark   Dividing by 2 here is just for the sake of an easy life: $\{\frac{1}{2}, \frac{1}{2}\}$ is a convenient solution of $\lambda + \mu = 1$. The point is just that $\lambda\mathbf{x} + \mu\mathbf{y}$ lies on a line containing chosen points of $0\mathbf{x}$ and $0\mathbf{y}$. The argument for (3) $\implies$ (2) can be made to work provided every line has ≥ 3 points, that is, over any field with > 2 elements.

Corollary   *A Euclidean motion $T : \mathbb{E}^n \to \mathbb{E}^n$ is an affine transformation, given in any choice of coordinates $\mathbb{E}^n \to \mathbb{R}^n$ by $T(\mathbf{x}) = A\mathbf{x} + \mathbf{b}$.*

This follows at once from Proposition 1.7, the implication (3) $\implies$ (1) in the previous proposition, and the fact that T is bijective, so the matrix A must be invertible.

1.10   Euclidean motions and orthogonal transformations

This section makes a brief use of the relationship between the standard quadratic form $|\mathbf{x}|^2 = \sum x_i^2$ on $\mathbb{R}^n$ and the associated inner product $\mathbf{x} \cdot \mathbf{y} = \sum x_i y_i$. If this is not familiar to you, I refer you once again to Appendix B for a general discussion.

Proposition   *Let A be an $n \times n$ matrix and $T : \mathbb{R}^n \to \mathbb{R}^n$ the map defined by $\mathbf{x} \mapsto A\mathbf{x}$. Then the following are equivalent conditions:*

(1) T is a motion $T : \mathbb{E}^n \to \mathbb{E}^n$.
(2) A preserves the quadratic form; that is, $|A\mathbf{x}| = |\mathbf{x}|$ for all $\mathbf{x} \in \mathbb{R}^n$.
(3) A is an orthogonal matrix; that is, it satisfies $^{t}AA = I_n$.

Proof   (1) $\implies$ (2) is trivial. Conversely,

$$|A\mathbf{x} - A\mathbf{y}|^2 = |A(\mathbf{x} - \mathbf{y})|^2 = |\mathbf{x} - \mathbf{y}|^2,$$

where the first equality is linearity, and the second follows from (2). Thus T preserves length, so it is a motion. (2) $\iff$ (3) is proved in Proposition B.4, where you can also read more about orthogonal matrixes if you wish to.   QED

Together with Corollary 1.7, this proves the following very important statement:

Corollary   *A Euclidean motion $T : \mathbb{E}^n \to \mathbb{E}^n$ is expressed in coordinates as*

$$T(\mathbf{x}) = A\mathbf{x} + \mathbf{b}$$

with A an orthogonal matrix, and $\mathbf{b} \in \mathbb{R}^n$ a vector.

An immediate check shows that an orthogonal matrix A has determinant $\det A = \pm 1$ (see Lemma B.4).

Definition   Let $T : \mathbb{E}^n \to \mathbb{E}^n$ be a motion expressed in coordinates as $T(\mathbf{x}) = A\mathbf{x} + \mathbf{b}$. I call T *direct* (or *orientation preserving*) if $\det A = 1$ and *opposite* (or *orientation reversing*) if $\det A = -1$.

The meaning of this notion in $\mathbb{E}^2$ and $\mathbb{E}^3$ is familiar in terms of left–right orientation, and it may seem pretty intuitive that it does not depend on the choice of coordinates. However, I leave the proof to Exercise 6.8.

1.11   Normal form of an orthogonal matrix

The point of this section is to express an orthogonal map $\alpha : \mathbb{R}^n \to \mathbb{R}^n$ in a simple form in a suitable orthonormal basis of $\mathbb{R}^n$. This section may seem an obscure digression into linear algebra, but the result is central to understanding motions of Euclidean space.

**1.11.1
The 2 × 2
rotation and
reflection
matrixes**

As a prelude to an attack on the general problem, consider the instructive case $n = 2$. The conditions for a 2×2 matrix $A = \left(\begin{smallmatrix} a & b \\ c & d \end{smallmatrix} \right)$ to be orthogonal are:

$$^{\mathrm{t}}AA = 1 \iff \begin{pmatrix} a & c \\ b & d \end{pmatrix} \begin{pmatrix} a & b \\ c & d \end{pmatrix} = \begin{pmatrix} 1 & 0 \\ 0 & 1 \end{pmatrix} \iff \begin{cases} a^2 + c^2 = 1 \\ ab + cd = 0 \\ b^2 + d^2 = 1. \end{cases}$$

Now $(a, c) \in \mathbb{R}^2$ is a point of the unit circle, so I can write $a = \cos\theta$, $c = \sin\theta$ for some $\theta \in [0, 2\pi)$ (Figure 1.11a). Then there are just two possibilities for b, d, giving

$$A = \begin{pmatrix} \cos\theta & -\sin\theta \\ \sin\theta & \cos\theta \end{pmatrix} \quad \text{or} \quad \begin{pmatrix} \cos\theta & \sin\theta \\ \sin\theta & -\cos\theta \end{pmatrix}.$$

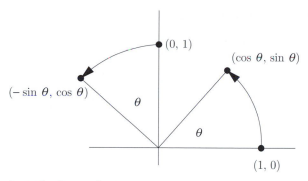

Figure 1.11a A rotation in coordinates.

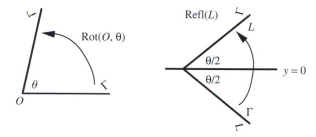

Figure 1.11b The rotation and the reflection.

The first of these corresponds to a direct motion (because $\det A = 1$), and you recognise it as a rotation around the origin through θ. In fact it takes

$$\begin{pmatrix} 1 \\ 0 \end{pmatrix} \mapsto \begin{pmatrix} \cos\theta \\ \sin\theta \end{pmatrix} \quad \text{and} \quad \begin{pmatrix} 0 \\ 1 \end{pmatrix} \mapsto \begin{pmatrix} -\sin\theta \\ \cos\theta \end{pmatrix}.$$

The second matrix gives an opposite motion ($\det A = -1$), and you can understand it in several ways; for example, write

$$A = \begin{pmatrix} \cos\theta & \sin\theta \\ \sin\theta & -\cos\theta \end{pmatrix} = \begin{pmatrix} \cos\theta & -\sin\theta \\ \sin\theta & \cos\theta \end{pmatrix} \begin{pmatrix} 1 & 0 \\ 0 & -1 \end{pmatrix}.$$

This says: first reflect in the x-axis, then rotate through θ. It is easy to see geometrically that this is the reflection in the line L through the origin 0 at angle $\theta/2$ to the x-axis. Indeed, every point on L is fixed, and the line perpendicular to L is reversed, as in Figure 1.11b.

In coordinates, this says that $\mathbf{f}_1 = (\cos(\theta/2), \sin(\theta/2))$ is an eigenvector of A with eigenvalue 1, and $\mathbf{f}_2 = (\sin(\theta/2), -\cos(\theta/2))$ an eigenvector with eigenvalue -1. The pair $(\mathbf{f}_1, \mathbf{f}_2)$ gives a vector space basis of $\mathbb{R}^2$, and in this new basis the map is given by the matrix $\begin{pmatrix} 1 & 0 \\ 0 & -1 \end{pmatrix}$. You can readily check these statements by matrix multiplication and the rules of trig, but the geometric argument is simpler and more convincing.

1.11.2   The general case   In the general case I control orthogonal matrixes using a slightly more involved argument.

> **Theorem (Normal form of orthogonal matrix)**   *Let $\alpha \colon \mathbb{R}^n \to \mathbb{R}^n$ be a linear map given by an orthogonal matrix A. Then there exists an orthonormal basis of $\mathbb{R}^n$ in which the matrix of α is*
>
> $$B = \begin{pmatrix} I_{k^+} & & & & & \\ & -I_{k^-} & & & & \\ & & B_1 & & & \\ & & & \ddots & \\ & & & & B_l \end{pmatrix} \quad where \quad B_i = \begin{pmatrix} \cos\theta_i & -\sin\theta_i \\ \sin\theta_i & \cos\theta_i \end{pmatrix}.$$
>
> *Here $k^+ + k^- + 2l = n$, and $I_{k^\pm}$ is the $k^\pm \times k^\pm$ identity matrix.*

Discussion   The rotation matrix $\left(\begin{smallmatrix} \cos\theta & -\sin\theta \\ \sin\theta & \cos\theta \end{smallmatrix} \right)$ has two special cases $\theta = 0$ (giving the identity) and $\theta = \pi$:

$$\begin{pmatrix} -1 & 0 \\ 0 & -1 \end{pmatrix} = \begin{pmatrix} \cos\pi & -\sin\pi \\ \sin\pi & \cos\pi \end{pmatrix} = 180° \text{ rotation.}$$

These trivial cases introduce a minor ambiguity in the normal form. The most natural convention seems to be to disallow $\theta = 0$, thus taking k^+ as big as possible, but to use $\theta = \pi$ wherever possible, so that $k^- = 0$ or 1.

Proof   In sketch form, this holds because A is orthogonal, so its eigenvalues have absolute value 1. Therefore they are either ± 1, or come in complex conjugate pairs $\{\lambda, \bar{\lambda}\} = \exp(\pm i\theta)$; after this, it is enough simply to build up a basis of $\mathbb{R}^n$ consisting either of real eigenvectors of A, or of real and imaginary parts of complex eigenvectors.

   Now I say the same thing again in more detail in 5 steps; the sketch proof just given already reveals that complex numbers are closely involved, so I may as well extend the action of A to the complex vector space $\mathbb{C}^n$, which I can do without any problems.

Step 1   If λ is a real eigenvalue of A then $\lambda = \pm 1$, because

$$A\mathbf{x} = \lambda\mathbf{x} \text{ and } A \text{ orthogonal} \implies |\mathbf{x}|^2 = |A\mathbf{x}|^2 = \lambda^2|\mathbf{x}|^2.$$

Step 2   If λ is a complex eigenvalue of A then $|\lambda| = 1$ and $\bar{\lambda} = \lambda^{-1}$ is also an eigenvalue (the bar denotes complex conjugate). Indeed, given $0 \neq \mathbf{z} \in \mathbb{C}^n$ such that $A\mathbf{z} = \lambda\mathbf{z}$ (recall I write $\mathbf{z} = {}^t(z_1, \dots, z_n)$ a column vector), write $\bar{\mathbf{z}} = {}^t(\bar{z}_1, \dots, \bar{z}_n)$.

Because A is a real matrix,

$$A\bar{\mathbf{z}} = \overline{A\mathbf{z}} = \overline{\lambda\mathbf{z}} = \bar{\lambda}\bar{\mathbf{z}}.$$

Now write $z_i = x_i + iy_i$, so that ${}^t\bar{\mathbf{z}}\mathbf{z} = \sum |z_i|^2 = \sum(x_i^2 + y_i^2) > 0$. Using the fact that A is orthogonal,

$$\bar{\lambda}\lambda\,{}^t\bar{\mathbf{z}}\mathbf{z} = {}^t(A\bar{\mathbf{z}})A\mathbf{z} = {}^t\bar{\mathbf{z}}\,{}^tAA\mathbf{z} = {}^t\bar{\mathbf{z}}\mathbf{z}, \quad \text{and thus} \quad \bar{\lambda}\lambda = 1.$$

Step 3   If $\lambda = \cos\theta + i\sin\theta$ is a complex eigenvalue of A (with $\theta \neq 0, \pi$) and $\mathbf{z} = \mathbf{x} + i\mathbf{y} \in \mathbb{C}^n$ a complex eigenvector then taking real and imaginary parts in the equality $A(\mathbf{x} + i\mathbf{y}) = A\mathbf{z} = \lambda\mathbf{z} = (\cos\theta + i\sin\theta)(\mathbf{x} + i\mathbf{y})$ gives

$$A\mathbf{x} = \cos\theta\mathbf{x} - \sin\theta\mathbf{y}, \quad A\mathbf{y} = \sin\theta\mathbf{x} + \cos\theta\mathbf{y}. \tag{10}$$

Now I claim that $|\mathbf{x}|^2 = |\mathbf{y}|^2$ and $\mathbf{x} \cdot \mathbf{y} = 0$, so that scaling makes $\mathbf{x}, \mathbf{y} \in \mathbb{R}^n$ into a pair of orthonormal vectors. This is an exercise for the reader. [Hint: write out the condition for (10) (with $\theta \neq 0, \pi$) to preserve $|\mathbf{x}|^2$, $|\mathbf{y}|^2$ and $\mathbf{x} \cdot \mathbf{y}$. See Exercises 1.5–1.6.]

Step 4   If α preserves a subspace W of $\mathbb{R}^n$, then it preserves its orthogonal complement under the inner product (compare B.3)

$$W^\perp = \{\mathbf{x} \in \mathbb{R}^n \mid \mathbf{x} \cdot \mathbf{w} = 0 \quad \text{for all } \mathbf{w} \in W\}.$$

In symbols,

$$\alpha(W) = W \implies \alpha(W^\perp) = W^\perp.$$

This is obvious from the definition of $W^\perp$. Look at Figure 1.15b for an example: if a motion preserves the horizontal plane W and its translates, then it will also preserve the orthogonal complement $W^\perp$, the vertical lines.

Step 5. Proof of the theorem   Eigenvalues of A come from the polynomial equation $p(\lambda) = \det(A - \lambda\mathbf{1}) = 0$, so that at least one real or complex eigenvalue λ exists. Step 1 or Steps 2–3 as appropriate gives a 1- or 2-dimensional subspace W with $AW = W$ on which the action of A is as indicated. By induction on the dimension, I can assume that the action of A on $W^\perp$ is OK; the induction starts with $\dim W = 0$ or 1.   QED

Complex numbers make their first incursion into real geometry during the above proof, and it is worth pondering why; quaternions also appear in a similar context in 8.5 below.

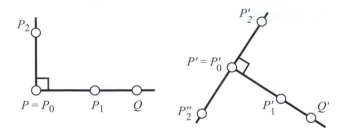

Figure 1.13   The Euclidean frames P_0, P_1, P_2 and P'_0, P'_1, P'_2.

1.12   Euclidean frames and motions

Definition   A *Euclidean frame* of $\mathbb{E}^n$ is a set of $n + 1$ points $Q_0, Q_1, \ldots, Q_n$ of $\mathbb{E}^n$ such that $d(Q_0, Q_i) = 1$ and the lines $Q_0 Q_i$ are pairwise orthogonal for $1 \leq i \leq n$.

Remark   The point of the definition is that if $Q_0, \ldots, Q_n$ is a Euclidean frame then it is possible to choose coordinates so that Q_0 becomes the origin $0 \in \mathbb{R}^n$ and the n vectors $\mathbf{e}_i = \overrightarrow{Q_0 Q_i}$ form an orthonormal basis of $\mathbb{R}^n$.

Theorem   *If we fix one Euclidean frame $P_0, P_1, \ldots, P_n$, then Euclidean motions are in one-to-one correspondence with Euclidean frames.*

Proof   The correspondence is given by $T \mapsto T(P_0), T(P_1), \ldots, T(P_n)$. It is clear that the image of the Euclidean frame $P_0, P_1, \ldots, P_n$ under a motion is again a Euclidean frame. The converse, that is, the fact that two Euclidean frames are mapped to each other by a unique motion, follows from the previous Remark and Appendix B, Proposition B.5.   QED

1.13   Frames and motions of $\mathbb{E}^2$

It is worth noting two useful consequences of Theorem 1.12, whose proofs are left as easy exercises (see Figure 1.13 and Exercise 1.12):

Corollary

(1)   *Suppose that $[P, Q]$ and $[P', Q']$ are two line segments in $\mathbb{E}^2$ of the same length $d(P, Q) = d(P', Q') > 0$. Then there exist exactly two motions $T : \mathbb{E}^2 \to \mathbb{E}^2$ such that $T(P) = P'$, and $T(Q) = Q'$.*

(2)   *Let $\triangle P Q R$ and $\triangle P' Q' R'$ be two triangles in $\mathbb{E}^2$ with all sides equal:*

$$d(P, Q) = d(P', Q'), \quad d(P, R) = d(P', R'), \quad d(Q, R) = d(Q', R').$$

(I assume that the three vertexes of each triangle are distinct and noncollinear.) Then there is a unique motion $T : \mathbb{E}^2 \to \mathbb{E}^2$ such that $T(P) = P', T(Q) = Q', T(R) = R'$.

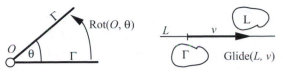

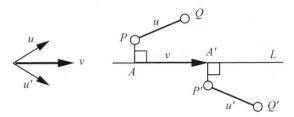

Figure 1.14a Rot(O, θ) and Glide(L, **v**).

Figure 1.14b Construction of glide.

1.14 Every motion of $\mathbb{E}^2$ is a translation, rotation, reflection or glide

Let us list the motions of $\mathbb{E}^2$ we know, expressed in coordinates (see Figure 1.14a).

1. The translation Trans(**b**): $\mathbf{x} \mapsto \mathbf{x} + \mathbf{b}$ for $\mathbf{b} \in \mathbb{R}^2$.
2. The rotation through angle θ about a point $O \in \mathbb{E}^2$; if O is the origin of the coordinate system, this is written

$$\text{Rot}(O, \theta): \begin{pmatrix} x_1 \\ x_2 \end{pmatrix} \mapsto \begin{pmatrix} \cos\theta & -\sin\theta \\ \sin\theta & \cos\theta \end{pmatrix} \begin{pmatrix} x_1 \\ x_2 \end{pmatrix}.$$

3. The reflection in a line L; if L is the x_1-axis ($x_2 = 0$) then

$$\text{Refl}(L): \begin{pmatrix} x_1 \\ x_2 \end{pmatrix} \mapsto \begin{pmatrix} x_1 \\ -x_2 \end{pmatrix}.$$

4. The *glide* (or glide reflection) in a line L through a vector **v** along L. Reflect in L and translate in **v**. If L is the x_1-axis ($x_2 = 0$) and $\mathbf{v} = (a, 0)$ then this is given by:

$$\text{Glide}(L, \mathbf{v}): \begin{pmatrix} x_1 \\ x_2 \end{pmatrix} \mapsto \begin{pmatrix} x_1 + a \\ -x_2 \end{pmatrix}.$$

Here **v** is parallel to L, and the reflection and translation commute.

I use self-documenting notation such as Rot(O, θ) and Glide(L, **v**) for these motions. In each case, I have chosen coordinates in an obvious way to make the formula as simple as possible. Obviously (1) and (2) are direct motions, and (3) and (4) opposite. Note that (3) is a particular case of (4) (where the translation vector is 0). It is sometimes convenient to view (1) as a limiting case of (2), when the centre of rotation is very far away and the angle of rotation correspondingly small.

Theorem   *That's all, folks!*

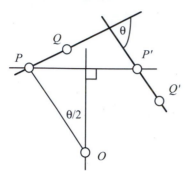

Figure 1.14c   Construction of rotation.

Proof   There are several ways of proving this. (Why not devise your own? See Exercises 1.8 and 1.9 for an argument in terms of $\mathbf{x} \mapsto A\mathbf{x} + \mathbf{b}$, and Exercise 2.11 for an argument in terms of composing reflections.)

The proof given here is based on the following geometric idea taken from Nikulin and Shafarevich [18]: let P, Q and P', Q' be two pairs of distinct points with $d(P, Q) = d(P', Q') \neq 0$. By Corollary 1.13, we know that there are *exactly two* motions of $\mathbb{E}^2$ such that $T(P) = P'$ and $T(Q) = Q'$. In Step 1 below, I construct a reflection or glide, and in Step 2 a rotation or translation. Now if T is any motion, pick any two distinct points $P \neq Q$, and set $P' = T(P)$, $Q' = T(Q)$. Then T must be one of the two motions constructed in Steps 1–2, both of which are in my list.

Step 1   I first find a reflection or glide. Write $\mathbf{u} = \overrightarrow{PQ}$ and $\mathbf{u}' = \overrightarrow{P'Q'}$. First I need to find the line of reflection L. The direction of L and of $\mathbf{v}$ is the vector bisecting the angle between $\mathbf{u}$ and $\mathbf{u}'$ (that is, $\frac{1}{2}(\mathbf{u} + \mathbf{u}')$ if the vectors are not opposite). Doing this arranges that the reflection or glide reflection in any line parallel to L takes $\overrightarrow{PQ}$ into a vector *parallel to* $\overrightarrow{P'Q'}$. Now choose L among lines with the given direction so that $d(L, P) = d(L, P')$, and write A and A' for the feet of the respective perpendiculars from P and P' to L and $\mathbf{v} = \overrightarrow{AA'}$ (see Figure 1.14b). Since reflection in L takes $\mathbf{u}$ into a vector parallel to $\mathbf{u}'$ by construction, and $d(P, Q) = d(P', Q')$, it is clear that Glide$(L, \mathbf{v})$ does what I want.

Step 2   There exists a rotation or translation $T : \mathbb{E}^2 \to \mathbb{E}^2$ such that $P \mapsto P'$ and $Q \mapsto Q'$. I suppose first that $P \neq P'$, and that the lines PQ and $P'Q'$ intersect at a single point in an angle θ.

Then the (signed) angle of rotation must be θ; the centre must be the point O of the perpendicular bisector of the line PP' determined by $POP' = \theta$ (see Figure 1.14c). Then by construction Rot(O, θ) takes $P \mapsto P'$, and the interval $[P, Q]$ to an interval out of P' of the same length with $d(P, Q) = d(P', Q')$ and the same direction as $[P'Q']$; hence it takes $Q \mapsto Q'$.

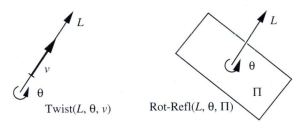

Figure 1.15a Twist (L, θ, **v**) and Rot-Refl (L, θ, Π).

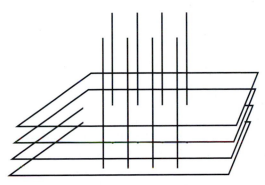

Figure 1.15b A grid of parallel planes and their orthogonal lines.

The proof just given does not work if $P = P'$, or if the lines PQ and $P'Q'$ are parallel, but these special cases are easy to deal with, and I leave them as exercises (see Exercise 1.10). QED

1.15 Classification of motions of $\mathbb{E}^3$

Theorem   *A motion $T : \mathbb{E}^3 \to \mathbb{E}^3$ is one of the following:*

1. Translation *by a vector* **v**.
2. Rotation *about a directed line L as axis through an angle θ.*
3. Twist: *the same followed by a translation along L (Figure 1.15a).*
4. Reflection *in a plane.*
5. Glide: *a reflection in a plane followed by a translation by a vector in the plane.*
6. Rotary reflection: *the rotation through θ about a directed axis L followed by a reflection in a plane Π perpendicular to L (Figure 1.15a).*

(2) is a special case of (3), and (4) is a special case of (5). In all cases where a motion is defined as a composite of two others, these two commute. (6) is also called a *rotary inversion*, because it is also the rotation around the directed axis L through $\pi + \theta$, followed by a point reflection in $L \cap \Pi$. Clearly (1)–(3) are direct motions and (4)–(6) opposite. Notice that any motion leaves invariant a grid of parallel planes and their orthogonal lines (Figure 1.15b).

Proof   See, for example, Exercise 1.11 or Rees [19], p. 16, Theorem 17 for a geometric proof. I give a coordinate geometry proof based on the use of the normal form of Theorem 1.11. Let $T : \mathbb{E}^3 \to \mathbb{E}^3$ be a motion expressed in coordinates as $T : \mathbf{x} \mapsto A\mathbf{x} + \mathbf{b}$; write $T = T_1 \circ T_2$ where T_i are given (in the same coordinate system) by

$$T_2 : \mathbf{x} \mapsto A\mathbf{x} \quad \text{and} \quad T_1 : \mathbf{y} \mapsto \mathbf{y} + \mathbf{b}.$$

Then by Theorem 1.11, there exists an orthogonal coordinate system such that

$$A = \begin{pmatrix} \pm 1 & \\ & B \end{pmatrix}, \quad \text{where} \quad B = \begin{pmatrix} \cos\theta & -\sin\theta \\ \sin\theta & \cos\theta \end{pmatrix}.$$

In these coordinates, T has the form

$$T : \begin{pmatrix} x_1 \\ x_2 \\ x_3 \end{pmatrix} \mapsto \left(\begin{matrix} \pm x_1 \\ \begin{pmatrix} \cos\theta & -\sin\theta \\ \sin\theta & \cos\theta \end{pmatrix} \begin{pmatrix} x_2 \\ x_3 \end{pmatrix} \end{matrix} \right) + \begin{pmatrix} b_1 \\ b_2 \\ b_3 \end{pmatrix}. \tag{11}$$

For the proof, I have to verify that this map is a motion of one of types (1)–(6). This can be done, for example, by a direct coordinate calculation. It is better to argue using the following *separation of variables*: (11) breaks T up as a *product* (not composite) of two motions $T = t' \times t'' : \mathbb{E}^1 \times \mathbb{E}^2 \to \mathbb{E}^1 \times \mathbb{E}^2$, where $T' : \mathbb{E}^1 \to \mathbb{E}^1$ and $T'' : \mathbb{E}^2 \to \mathbb{E}^2$ are given in coordinates by

$$T' : x_1 \mapsto \pm x_1 + b_1 \quad \text{and} \quad T'' : \begin{pmatrix} x_2 \\ x_3 \end{pmatrix} \mapsto \begin{pmatrix} \cos\theta & -\sin\theta \\ \sin\theta & \cos\theta \end{pmatrix} \begin{pmatrix} x_2 \\ x_3 \end{pmatrix} + \begin{pmatrix} b_2 \\ b_3 \end{pmatrix}.$$

In other words, (11) separates the 3 variables in such a way that $T(\mathbf{x}) = \mathbf{y}$ with $\mathbf{y} = (y_1, y_2, y_3)$, where y_1 is a function of x_1 only, and y_2, y_3 functions of x_2, x_3 only. Now both T' and T'' are motions in their own right. This is the real point of the theorem. (It is easy to generalise the result to all dimensions; compare Theorem 2.5.)

T'' is a direct motion, and is a translation if $\theta = 0$ or rotation if $\theta \neq 0$; this follows by Theorem 1.14, or by direct observation. In terms of coordinates (x_2, x_3) of $\mathbb{E}^2$, it is the rotation through an angle θ about the point determined by

$$\begin{pmatrix} x_2 \\ x_3 \end{pmatrix} = \begin{pmatrix} \cos\theta & -\sin\theta \\ \sin\theta & \cos\theta \end{pmatrix} \begin{pmatrix} x_2 \\ x_3 \end{pmatrix} + \begin{pmatrix} b_2 \\ b_3 \end{pmatrix},$$

that is, solving for x_2, x_3 by inverting a 2×2 matrix:

$$\begin{pmatrix} x_2 \\ x_3 \end{pmatrix} = \frac{-1}{2 - 2\cos\theta} \begin{pmatrix} \cos\theta - 1 & -\sin\theta \\ \sin\theta & \cos\theta - 1 \end{pmatrix} \begin{pmatrix} b_2 \\ b_3 \end{pmatrix}.$$

The theorem follows easily on sorting out the cases.   QED

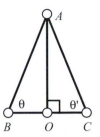

Figure 1.16a Pons asinorum.

1.16 Sample theorems of Euclidean geometry

This chapter has mainly been concerned with the foundations of Euclidean geometry and a description of Euclidean motions. I do not have time to give many results of substance from Euclidean geometry, either the theory of Euclid's Elements, or the much more extensive nineteenth century subject, but I do not want to omit to mention it altogether. Coxeter [5] is very entertaining on this subject.

**1.16.1
Pons asinorum**

Proposition Pons asinorum, '*Bridge of asses*'. *Equivalent conditions on a triangle* $\triangle ABC$:

1. $d(A, B) = d(A, C)$;
2. $\theta = \angle ABC = \theta' = \angle ACB$;
3. *there exists a motion* $T: \triangle ABC \mapsto \triangle ACB$.

Proof (1) $\iff$ (2) is an easy consequence of trigonometry, because in Figure 1.16a,

$$d(A, O) = d(A, B) \sin \theta = d(A, C) \sin \theta'.$$

From our point of view, (3) $\implies$ (1) or (2) is obvious, and (1) or (2) $\implies$ (3) follows by Corollary 1.13. You can also directly invoke the motion of the plane consisting of picking up the triangle and laying it down over itself so that A, B, C match up with A, C, B in order; alternatively, you can drop a perpendicular AO from A to BC, and argue on congruent triangles. QED

**1.16.2
The angle sum of triangles**

Theorem *The sum of angles in a triangle is equal to π.*

Proof Let $\triangle ABC$ be a given triangle. Consider the motion $T = \text{Trans}(\overrightarrow{AC})$ and set $\triangle A'B'C' = T(\triangle ABC)$ as in Figure 1.16b. Then because T is a motion, I get $\triangle A'B'C' \equiv \triangle ABC$, where $\equiv$ is congruence (see Exercise 1.16). Also, since T is a Euclidean translation, $d(B, B') = d(A, C)$, therefore also $\triangle ABC \equiv \triangle BA'B'$. Hence

$$\alpha + \beta + \gamma = \angle B'CC' + \angle BCB' + \angle ACB = \pi$$

since the angles combine to form a straight line. QED

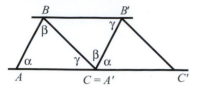

Figure 1.16b   Sum of angles in a triangle is equal to π.

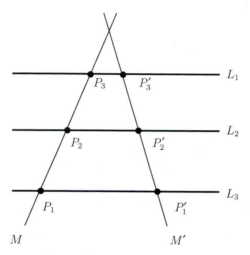

Figure 1.16c   Parallel lines fall on lines in the same ratio.

Remark   The statement *the sum of angles in a triangle equals π* is equivalent to the parallel postulate (see 3.13 and 9.1.2). The proof used translation in $\mathbb{E}^2$, coming from the coordinate model. Figure 1.16b makes sense in spherical geometry (or hyperbolic geometry), but there $d(A, A') > d(B, B')$ (respectively $d(A, A') < d(B, B')$).

**1.16.3
Parallel
lines and
similar
triangles**

A distinguishing feature of Euclidean geometry is the existence of unique parallel lines (compare 9.1.2). Parallel lines fall on lines in the same ratio, and conversely; they are also responsible for the existence of similar triangles. The following proposition makes these statements precise.

Proposition

(1)   If L_1, L_2, L_3 are three parallel lines in $\mathbb{E}^2$, and they meet a line M in P_1, P_2, P_3, then the (signed) ratio of distances $d(P_1, P_2) : d(P_2, P_3)$ is independent of M (Figure 1.16c).

(2)   Consider the two triangles $\triangle ABC$ and $\triangle AB'C'$ of Figure 1.16d. The following are equivalent:

(a) BC is parallel to $B'C'$.
(b)   Equality of ratios: $d(A, B) : d(A, B') = d(A, C) : d(A, C')$.
(c)   Equality of angles: $\angle ABC = \angle AB'C'$ and $\angle ACB = \angle AC'B'$.

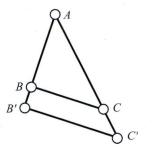

Figure 1.16d Similar triangles.

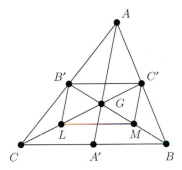

Figure 1.16e The centroid.

Proof   All this is trivial in coordinate geometry; see Exercise 1.17.

Two triangles satisfying the conditions of the second part are called *similar*. Corresponding pairs of angles of a pair of similar triangles are equal.

1.16.4
Four centres
of a triangle

Proposition (Centroid)   *The three medians of a triangle ABC meet in a point G (Figure 1.16e).*

Proof   (See 4.7 for a slightly different proof.) Let A', B', C' be the midpoints of BC, AC, AB and let G be the point on AA' with $d(A, G) = 2d(G, A')$. If L, M are the midpoints of AG and CG, then by similar triangles

$$LM \parallel AC \parallel A'C' \quad \text{and} \quad LC' \parallel GB \parallel MA',$$

(where $\parallel$ is parallel), so that $LMA'C'$ is a parallelogram, G is its centre, so MGC' is a straight line. Hence G lies on each of AA', BB', CC', so it is the centroid.   QED

Proposition (Circumcentre)   *The three perpendicular bisectors of sides AB, BC and AC meet in a point O. This is the centre of the circle circumscribed around ABC (Figure 1.16f).*

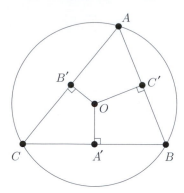

Figure 1.16f   The circumcentre.

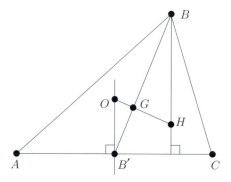

Figure 1.16g   The orthocentre.

Proof   This is almost obvious, since the perpendicular bisector of AB is determined as the locus of points equidistant from A and B, so that any two of the perpendicular bisectors intersect at the point O determined by $d(A, O) = d(B, O) = d(C, O)$.   QED

Proposition (Orthocentre)   *The three perpendiculars dropped from a vertex onto the opposite side intersect in a point H.*

Proof   In vector notation, H is the point given by $\overrightarrow{OH} = 3\overrightarrow{OG}$, where O is the circumcentre and G the centroid. Indeed, in Figure 1.16g, BB' is the median and OB' the perpendicular bisector of AC; since $\overrightarrow{GB} = 2\overrightarrow{B'G}$ and $\overrightarrow{GH} = 2\overrightarrow{OG}$, it follows that the two triangles $\triangle GB'O$ and $\triangle GBH$ are similar. Therefore the line BH is perpendicular to AC, and H lies on this perpendicular. H lies on each of the other two perpendiculars for similar reasons.   QED

Note that, as a byproduct of the above proof, we also see that the centroid G lies on the segment $[O, H]$ determined by the circumcentre and the orthocentre, and divides it into the ratio $(1 : 2)$.

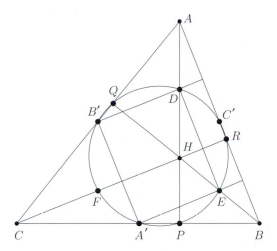

Figure 1.16h The Feuerbach 9-point circle.

Proposition (Incentre)   *The angle bisectors of the three angles $\angle CAB$, $\angle ABC$ and $\angle ACB$ meet in a point K. This is the centre of the circle inscribed into ABC.*

Proof   This is exactly analogous to the case of the circumcentre above (see Exercise 1.18).   QED

1.16.5
The
Feuerbach
9-point
circle

Theorem (The Feuerbach circle)[1]   *The following 9 points lie on a circle (see Figure 1.16h):*

3 feet P, Q, R of the perpendiculars dropped from a vertex to the opposite side;
3 midpoints A', B', C' of the sides;
3 midpoints D, E, F of AH, BH, CH, where H is the orthocentre.

Proof   The intellectual achievement here is the statement, of course. The proof is rather easy because there are so many parallel and perpendicular lines in Figure 1.16h. By similar triangles, the following lines are parallel:

$$A'B' \parallel DE \parallel AB \quad \text{and} \quad A'E \parallel B'D \parallel CR.$$

But $AB \perp CR$ by construction, hence $A'B'DE$ is a rectangle. Thus the circle with diameter $A'D$ also has $B'E$ as diameter; arguing in the same way one sees that $A'C'DF$ is also a rectangle, so that the same circle with diameter $A'D$ also has $C'F$ as diameter. Finally, $\angle A'PD = 90°$, which is a sufficient condition for the circle with diameter $A'D$ to pass through P, so that this same circle passes also through the feet of the perpendiculars.   QED

---

[1] The Feuerbach circle is alternatively called the Euler circle, because it was discovered by Poncelet and Brianchon. The reason why the young Bavarian schoolmaster Feuerbach's name appears in the context is his beautiful theorem that the circle touches the inscribed circle of the triangle. Purists may prefer the noncommital name 9-point circle.

Exercises

1.1 Redo the proof of Theorem 1.1 in detail in the cases $n = 1$ and $n = 2$.

1.2 The angle between nonzero vectors $\mathbf{u}, \mathbf{v} \in \mathbb{R}^n$ can be defined by

$$\cos \theta = \sum u_i v_i / |\mathbf{u}||\mathbf{v}|.$$

Prove that the right-hand side is in the interval $[-1, +1]$, so that its arccos is defined.

1.3 The line $L = \mathbf{xy}$ in $\mathbb{R}^n$ is the set $\{(1 - \lambda)\mathbf{x} + \lambda\mathbf{y}|\lambda \in \mathbb{R}\}$. If $\mathbf{z} \in L$, write $\mathbf{y}$ in terms of $\mathbf{x}$ and $\mathbf{z}$. Complete the proof of Proposition 1.8.

1.4 Show that the assumption that T is bijective in the definition of motion of Euclidean space is superfluous; that is, a map $T : \mathbb{E}^n \to \mathbb{E}^n$ that preserves distances is bijective, therefore a motion. [Hint: prove that T is affine linear. Compare Exercise A.1.]

1.5 Complete the proof of Step 3 in Theorem 1.11 using the hint given in the text.

1.6 Let A be a (real) orthogonal matrix.

(a) If $\mathbf{e}, \mathbf{f} \in \mathbb{R}^n$ are eigenvectors of A belonging to distinct eigenvalues $\lambda \neq \mu$, prove that $\mathbf{e} \cdot \mathbf{f} = 0$.

(b) If $\mathbf{z} \in \mathbb{C}^n$ is a complex eigenvector with complex eigenvalue $\lambda \notin \mathbb{R}$, prove that $\mathbf{z} \cdot \mathbf{z} = 0$. (Here $\mathbf{x} \cdot \mathbf{y} = \sum_j x_j y_j$ is the usual inner product.) Use this to give a better proof of Step 3 in Theorem 1.11.

1.7 (a) Let $T : \mathbb{E}^2 \to \mathbb{E}^2$ be the motion obtained by reflecting in the x-axis then rotating through θ around the origin. Show that T is the reflection in a certain line (to be specified).

(b) Calculate the eigenvalues and eigenvectors of the reflection matrix $A = \begin{pmatrix} \cos\theta & \sin\theta \\ \sin\theta & -\cos\theta \end{pmatrix}$.

(c) Relate (a) and (b).

1.8 (a) Let θ be a nonzero angle and $\mathbf{b}$ a translation vector in the plane. Give a geometric construction for a point $P \in \mathbb{E}^2$ such that

$$\text{Rot}(O, \theta)(P) = \text{Trans}(-\mathbf{b})(P).$$

[Hint: draw a picture, to find points P, Q with $\mathbf{b} = \overrightarrow{QP}$ such that O is on the perpendicular bisector of PQ and $\angle POQ = \theta$.]

(b) By solving linear equations, find x, y such that

$$A\begin{pmatrix} x_1 \\ x_2 \end{pmatrix} + \begin{pmatrix} b_1 \\ b_2 \end{pmatrix} = \begin{pmatrix} x_1 \\ x_2 \end{pmatrix}, \quad \text{where} \quad A = \begin{pmatrix} \cos\theta & \sin\theta \\ \sin\theta & -\cos\theta \end{pmatrix}.$$

(c) Express the motion $T : \mathbb{E}^2 \to \mathbb{E}^2$ defined in coordinates by $T(\mathbf{x}) = A\mathbf{x} + \mathbf{b}$ in the form $T = \text{Rot}(P, \theta)$.

(d) Relate (a) and (b).

1.9 Let $A = \begin{pmatrix} \cos\theta & \sin\theta \\ \sin\theta & -\cos\theta \end{pmatrix}$ be the reflection matrix of 1.11.1, and consider the motion $T(\mathbf{x}) = A\mathbf{x} + \mathbf{b}$; give a proof in coordinates that it is a glide reflection. [Hint: you need to turn Figure 1.14b into coordinates.]

1.10 In the proof of Theorem 1.14, Step 2, there are 3 special cases:

(a) $P = P'$,

(b) PQ and $P'Q'$ are parallel,

(c) and PQ and $P'Q'$ are opposite (that is PQ and $Q'P'$ parallel).

Complete the proof of Step 2 in any of these cases by constructing a suitable translation or rotation taking $P \mapsto P'$ and $Q \mapsto Q'$.

1.11 Find the two motions $\mathbb{E}^2 \to \mathbb{E}^2$ taking $(0, 0) \mapsto (1, 2)$ and $(0, \sqrt{2}) \mapsto (2, 3)$. Write each as $\mathbf{x} \mapsto A\mathbf{x} + \mathbf{b}$. [Hint: the easy way: for the direct motion, translate then rotate; for the opposite motion, reflect then translate then rotate.] Express them as rotation and glide.

1.12 Prove Corollary 1.13 (1). [Hint: as in Figure 1.13, make a Euclidean frame with $P_0 = P$, $\overrightarrow{P_0 P_1} = \frac{\overrightarrow{PQ}}{d(P,Q)}$ and P_2 a third point; if I do the same for P', Q', there are 2 choices for P_2', one on either side of the line $P'Q'$. The statement now follows by Theorem 1.12.]

1.13 Let $P_0, P_1, P_2 \in \mathbb{E}^2$ be distinct noncollinear points. Show that there is a unique Euclidean frame so that $P_0 = (0, 0)$, $P_1 = (a, 0)$ with $a > 0$ and $P_2 = (b, c)$ with $c > 0$. Deduce that a motion of $\mathbb{E}^2$ is uniquely determined by its effect on any 3 distinct noncollinear points.

1.14 Let P_0, P_1, P_2 and $P_0', P_1', P_2' \in \mathbb{E}^2$ be two pairs of distinct noncollinear points such that $d(P_i, P_j) = d(P_i', P_j')$ for all i, j. Prove that there exists a unique motion $T : \mathbb{E}^2 \to \mathbb{E}^2$ taking $P_i \mapsto P_i'$ for $i = 1, 2, 3$. [Hint: you know enough motions to send $P_0 \mapsto P_0'$. Then fixing $P_0 = P_0'$, to send $P_1 \mapsto P_1'$ in exactly 2 different ways. Where does this leave P_2?]

1.15 Let $P_0, \ldots, P_n$ be $n + 1$ points spanning $\mathbb{E}^n$. Prove that a point $Q \in \mathbb{E}^n$ is uniquely determined by its distances from all of the P_i. [Hint: take P_0 as origin; the n vectors $\mathbf{e}_i = \overrightarrow{P_0 P_i}$ are linearly independent. The vector $\mathbf{f} = \overrightarrow{P_0 Q}$ is determined by $\mathbf{f} \cdot \mathbf{e}_i$, so it is enough to determine $\mathbf{f} \cdot \mathbf{e}_i$ from distances in $\triangle P_0 P_i Q$.]

1.16 Let $\triangle ABC$ and $\triangle DEF$ be two triangles in $\mathbb{E}^2$. Prove that the following 4 conditions are equivalent:
   (a) 3 sides are equal $AB = DE$, $BC = EF$, $CA = FD$;
   (b) equal side–angle–side: $AB = DE$, $CA = FD$ and $\angle CAB = \angle FDE$;
   (c) angle–side–angle: $\angle ABC = \angle DEF$, $BC = EF$ and $\angle BCA = \angle EFD$;
   (d) there exists a motion T taking $A \mapsto D$, $B \mapsto E$, $C \mapsto F$.
   The triangles $\triangle ABC$ and $\triangle DEF$ are *congruent* if these conditions hold; in symbols, $\triangle ABC \equiv \triangle DEF$.

1.17 Prove Proposition 1.16.3 by computing in a suitably chosen coordinate system.

1.18 By analogy with the proof of Proposition 1.16.4 (Circumcentre), prove that the three angle bisectors of angles $\angle CAB$, $\angle ABC$ and $\angle ACB$ meet in a point K. Show also that this is the centre of the circle inscribed in ABC (a circle touching all sides of $\triangle ABC$).

2 Composing maps

This brief chapter takes up some examples and simple applications of composition of maps. The aim is to clarify and review some results about motions from Chapter 1, and to prepare some foundational points for later chapters. Composing maps is the idea of taking 'a function of a function', a procedure familiar from first year calculus: if $y = f(x)$ and $z = g(y)$, then you can write $z = g(f(x)) = (g \circ f)(x)$. The chain rule, for example, calculates the derivative $\frac{dz}{dx}$ in terms of $\frac{dy}{dx}$ and $\frac{dz}{dy}$.

2.1 Composition is the basic operation

One may consider the fundamental objects in math to be numbers of various kinds; the basic operations on them are then addition and multiplication (together with subtraction, division, taking roots, etc., which are in some sense the inverses of the basic operations). There would be no point in having numbers if you could not calculate with them. The reason that we use numbers to model the real world is precisely that it is easier to perform operations on numbers than make the corresponding constructions on objects out there in the wild.

However, at another level, the fundamental objects might be maps between sets. Then the basic operation is *composition of maps*. Let X, Y, Z be sets, and $f : X \to Y$ and $g : Y \to Z$ two maps between them.

Definition   The *composite* of f and g is the map

$$g \circ f : X \to Z \quad \text{defined by} \quad (g \circ f)(x) = g(f(x)). \tag{1}$$

This may look like an associative law – but in reality it is just the definition of the left-hand side. The left-hand side is pronounced 'g follows f, applied to x'.

The first point is that composition is a basic operation, comparable to addition and multiplication of numbers.

1. Composing two translations of $\mathbb{E}^n$ means adding the corresponding vectors:

$$\text{Trans}(\mathbf{v}) \circ \text{Trans}(\mathbf{u}) = \text{Trans}(\mathbf{u} + \mathbf{v}).$$

Indeed, either side is the operation $\mathbf{x} \mapsto \mathbf{x} + \mathbf{u} + \mathbf{v}$.

2. Composing two rotations of $\mathbb{E}^2$ (about the same centre) means adding the corresponding angles (modulo 2π):

$$\text{Rot}(\theta) \circ \text{Rot}(\varphi) = \text{Rot}(\theta + \varphi).$$

This is clear if you draw the picture; it gives the identity

$$\begin{pmatrix} \cos\theta & -\sin\theta \\ \sin\theta & \cos\theta \end{pmatrix} \begin{pmatrix} \cos\varphi & -\sin\varphi \\ \sin\varphi & \cos\varphi \end{pmatrix} = \begin{pmatrix} \cos(\theta + \varphi) & -\sin(\theta + \varphi) \\ \sin(\theta + \varphi) & \cos(\theta + \varphi) \end{pmatrix}.$$

3. In linear algebra, a matrix corresponds to a linear map; the product of two matrixes is the composite of the corresponding linear maps (see Exercise 2.1).

4. One way to introduce complex numbers is as similarities of $\mathbb{E}^2$: a complex number $z = r\exp(i\theta)$ corresponds to rotation by θ together with a dilation by a factor r. In these terms, product of complex numbers is composite of maps (see Exercise 2.2).

2.2 Composition of affine linear maps $\mathbf{x} \mapsto A\mathbf{x} + \mathbf{b}$

An affine linear map $T : \mathbb{R}^n \to \mathbb{R}^n$ is given by $T(\mathbf{x}) = A\mathbf{x} + \mathbf{b}$ where A is an $n \times n$ matrix and $\mathbf{b}$ is a vector (see 1.8). If $T_1(\mathbf{x}) = A_1\mathbf{x} + \mathbf{b}_1$ and $T_2(\mathbf{x}) = A_2\mathbf{x} + \mathbf{b}_2$ then

$$(T_2 \circ T_1)(\mathbf{x}) = A_2 T_1(\mathbf{x}) + \mathbf{b}_2 = A_2(A_1\mathbf{x} + \mathbf{b}_1) + \mathbf{b}_2 = (A_2 A_1)\mathbf{x} + (A_2\mathbf{b}_1 + \mathbf{b}_2).$$

Thus if we write $T_{A,\mathbf{b}}$ for the map $\mathbf{x} \mapsto A\mathbf{x} + \mathbf{b}$, composition is given by the rule $T_{A_2,\mathbf{b}_2} \circ T_{A_1,\mathbf{b}_1} = T_{A_2 A_1, A_2\mathbf{b}_1 + \mathbf{b}_2}$. Note that the first component $A_2 A_1$ is just the product, whereas in the second component, the matrix A_2 of $T_{A_2,\mathbf{b}_2}$ first acts on the translation vector $\mathbf{b}_1$ before the vectors are added. I return to this composition rule in 6.5.3 below; compare also Exercise 6.1.

2.3 Composition of two reflections of $\mathbb{E}^2$

Consider the reflections of $\mathbb{E}^2$ in two lines L_1, L_2. There are two cases (see Figure 2.3):

1. If L_1 and L_2 meet in a point P and θ is the angle at P from L_1 to L_2 then $\text{Refl}(L_2) \circ \text{Refl}(L_1) = \text{Rot}(P, 2\theta)$.

2. If L_1 and L_2 are parallel and $\mathbf{v}$ is the perpendicular vector from L_1 to L_2 then $\text{Refl}(L_2) \circ \text{Refl}(L_1) = \text{Trans}(2\mathbf{v})$.

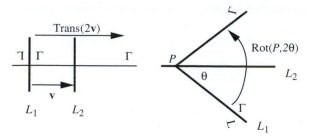

Figure 2.3   Composite of two reflections.

2.4   Composition of maps is associative

I want to consider the composite of many maps in what follows, for example the composite of 3 reflections $\mathrm{Refl}(L_3) \circ \mathrm{Refl}(L_2) \circ \mathrm{Refl}(L_1)$. As a preliminary step, a point of set theory: suppose that X, Y, Z, T are sets, and that

$$f : X \to Y, \quad g : Y \to Z, \quad h : Z \to T$$

are three maps. The associative law is the tautology that *there is only one way of getting from X to T using f, g, h in that order*, namely

$$x \mapsto f(x) \mapsto g(f(x)) \mapsto h(g(f(x))). \tag{2}$$

The composite $h \circ g \circ f$ is the map $X \to T$ defined by (2). Thus the expression $h \circ g \circ f$ does not admit any possible ambiguity.

In the tradition of abstract algebra, the associative law is the headache of how to bracket $h \circ g \circ f$. It occurs if we think of the composite of only two maps as the basic operation, and interpret a composite of three or more maps in a recursive way, such as $h \circ (g \circ f)$, presumably to economise on definitions. In this case, one first constructs a map $g \circ f : X \to Z$, then links it with the third map to get the repeated composite $h \circ (g \circ f) : X \to Z \to T$. However, as my tautology says, whatever brackets you put in, $h \circ g \circ f$ has only one possible meaning, namely (2). You can think through a few of these identities as exercises, see Exercise 2.3. (I warn you, it is exceedingly boring.)

Another abstract algebraic notion, the 'commutative law', is discussed in Exercise 2.4.

2.5   Decomposing motions

This section introduces the first way of decomposing a motion of $\mathbb{E}^n$ as a composite of 'elementary' motions. Although there are more powerful decompositions around (see for example the next section), the one given here already illustrates some basic features of any such decomposition. To start with, let us make a list of motions of $\mathbb{E}^n$ that could reasonably be called 'elementary'.

An affine linear subspace $\Sigma \subset \mathbb{E}^n$ of Euclidean space is the image $U \subset \mathbb{R}^n$ of a vector subspace under some choice of coordinates. The dimension of Σ is the dimension $\dim U$ of U. (These notions will be investigated in much more detail in 4.3 below.) In particular, a *hyperplane* of $\mathbb{E}^n$ is an $(n-1)$-dimensional affine linear subspace $\Pi \subset \mathbb{E}^n$.

Definition The *reflection* in a hyperplane Π is the motion that fixes Π pointwise and reverses orthogonal vectors to Π. In coordinate form, if Π is given by $x_1 = 0$, and $x_2, \ldots, x_n$ are coordinates on $\Pi \cong \mathbb{E}^{n-1}$, then

$$\text{Refl}(\Pi): \begin{pmatrix} x_1 \\ \vdots \\ x_n \end{pmatrix} \mapsto \begin{pmatrix} -1 & & & \\ & 1 & & \\ & & \ddots & \\ & & & 1 \end{pmatrix} \begin{pmatrix} x_1 \\ \vdots \\ x_n \end{pmatrix}.$$

In other words, the defining property of $\rho = \text{Refl}(\Pi)$ is that it fixes every point of Π, and takes $P \notin \Pi$ into the point $Q = \rho(P)$ such that Π is the perpendicular bisector of PQ. Note that if P and Q are two distinct points of $\mathbb{E}^n$, there is a unique hyperplane Π such that $\text{Refl}(\Pi)$ takes P to Q, namely the perpendicular bisector of PQ; this is also determined as the locus of points equidistant from P and Q.

Definition Let Σ be an $(n-2)$-dimensional affine linear subspace of $\mathbb{E}^n$. The *rotation* around the axis Σ through (signed) angle θ is the motion that fixes Σ pointwise and rotates by θ in planes orthogonal to Σ.

In coordinates, if Σ is given by $x_1 = x_2 = 0$, then the planes orthogonal to Σ are described by $x_3 = c_3, \ldots, x_n = c_n$ for $c_3, \ldots, c_n$ real constants (draw a picture for $n = 3$!). Hence the coordinate form is

$$\text{Rot}(\Sigma, \theta): \begin{pmatrix} x_1 \\ \vdots \\ x_n \end{pmatrix} \mapsto \begin{pmatrix} \cos\theta & -\sin\theta & & & \\ \sin\theta & \cos\theta & & & \\ & & 1 & & \\ & & & \ddots & \\ & & & & 1 \end{pmatrix} \begin{pmatrix} x_1 \\ \vdots \\ x_n \end{pmatrix}.$$

Finally, there are also *translations* $\text{Trans}(\mathbf{v}): \mathbf{x} \mapsto \mathbf{x} + \mathbf{b}$ for $\mathbf{b} \in \mathbb{R}^n$.

Theorem *Every motion T of $\mathbb{E}^n$ is a composite of a translation, k reflections and l rotations, where $k + 2l \leq n$.*

Proof Convince yourself that this is really a restatement of the fact that every orthogonal matrix has a normal form described in Theorem 1.11. QED

2.6 Reflections generate all motions

Here we aim to improve the statement of the previous section, using geometric rather than algebraic reasoning.

Theorem   *Every motion T of $\mathbb{E}^n$ is a composite of at most $n + 1$ reflections,*

$$T = \rho_1 \circ \rho_2 \circ \cdots \circ \rho_k, \quad \text{with } k \leq n + 1.$$

Proof   The rough idea is simple: if every point $P \in \mathbb{E}^n$ is fixed by T, then $T =$ id, so it is a composite of no reflections at all. Otherwise, choose any P so that $T(P) = Q \neq P$; then, by what I just said, there is a reflection ρ_1 taking Q back to P, namely the reflection in the perpendicular bisector of PQ. Then $T(P) = Q$ and $\rho_1(Q) = P$, so that $T_1 = \rho_1 \circ T$ is a new motion fixing P. Now it turns out (see below) that T_1 still fixes any point already fixed by T, so that T_1 fixes strictly more than T. I can repeat this argument, obtaining $T_2 = \rho_2 \circ T_1$ fixing even more points, and so on inductively until $T_k = \rho_k \circ T_{k-1}$ fixes every point of $\mathbb{E}^n$. Putting this together gives $\rho_k \circ \cdots \circ \rho_1 \circ T =$ id.

Now precomposing the equation $T_1 = \rho_1 \circ t$ with ρ_1 gives

$$\rho_1 \circ T_1 = (\rho_1 \circ \rho_1) \circ T,$$

and since $\rho_1 \circ \rho_1 =$ id, we get $T = \rho_1 \circ T_1$. Arguing in the same way gives $T = \rho_1 \circ T_1 = \rho_1 \circ \rho_2 \circ T_2 = \cdots$, which concludes the proof.

To go through the argument in more detail, I assert first that the set $\mathrm{Fix}(T)$ of fixed points of any motion T is (either empty or) an affine linear subspace of $\mathbb{E}^n$. This follows from Proposition 4.3 (2), and the fact that if two distinct points P, Q are fixed by T, then so is any point R on the line PQ: if $R \in [P, Q]$ then

$$d(P, R) + d(R, Q) = d(P, Q) \quad \text{and} \quad T(P) = P, \ T(Q) = Q$$

$$\implies d(P, T(R)) + d(T(R), Q) = d(P, Q),$$

so $T(R) \in [P, Q]$ and $T(R) = R$, and similarly if P, Q, R are collinear but in some other order.

Now to get a neat induction, I add a slightly stronger clause to the theorem:

Claim   *Moreover, if $\mathrm{Fix}(T)$ has dimension $n - l$ (for some $l = 0, \ldots, n$) then T is a composite of at most l reflections.*

As argued above, if $T \neq$ id then I choose a point $P \notin \mathrm{Fix}(T)$, set $Q = T(P)$ and Π the perpendicular bisector of PQ, and let ρ be the reflection in Π. The point of the construction is that $\rho(Q) = P$, so that $T_1 = \rho \circ T$ fixes P.

Now the perpendicular bisector Π is characterised as the set of points of $\mathbb{E}^n$ equidistant from P and Q. Moreover, every point $R \in \mathrm{Fix}(T)$ is equidistant from P and Q, because $d(P, R) = d(T(P), T(R)) = d(Q, R)$. Therefore $\mathrm{Fix}(T) \subset \Pi$, and $\rho = \mathrm{Refl}(\Pi)$ fixes every point of $\mathrm{Fix}(T)$. It follows that $\mathrm{Fix}(T_1) \supset \mathrm{Fix}(T) \cup \{P\}$.

The claim now follows by induction on l. If $l = 0$ then $T =$ id. If $l = 1$ then $\mathrm{Fix}(T) = \Pi$ is a hyperplane, and $T = \mathrm{Refl}(\Pi)$. Otherwise, as just proved, I can find ρ so that $T_1 = \rho \circ T$ fixes a strictly bigger set than T, and therefore $\mathrm{Fix}(T_1)$ has

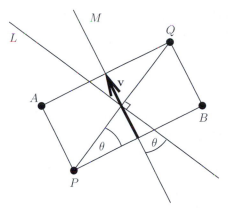

Figure 2.7   Composite of a rotation and a reflection.

dimension $(n - l')$ with $l' < l$. By induction, I can assume the result for T_1, that is, $T_1 = \rho_1 \circ \rho_2 \circ \cdots \circ \rho_k$ with $k \le l'$ so that $T = \rho \circ T_1$ is the composite of at most $l' + 1 \le l$ reflections, as required. This proves the claim. If $\mathrm{Fix}(T) = \emptyset$ then $\mathrm{Fix}(T_1)$ is at least one point, so that by the claim, T_1 is a composite of at most n reflections, and T the composite of at most $n + 1$ reflections, which proves the theorem.   QED

2.7 An alternative proof of Theorem 1.14

Theorem (= Theorem 1.14)   *Every motion of $\mathbb{E}^2$ is a rotation, reflection, translation or a glide.*

Proof   Every motion of $\mathbb{E}^2$ is the composite of at most 3 reflections. As we saw in 2.3, the composite of 2 reflections is a translation if the 2 axes are parallel, and a rotation if they meet at a point P. It only remains to prove that the composite of 3 reflections $\rho_3 \circ \rho_2 \circ \rho_1$ is a glide or reflection. Suppose for simplicity that the axes of ρ_1 and ρ_2 meet at a point P, and make an angle θ there, so that $\rho_2 \circ \rho_1 = \mathrm{Rot}(P, 2\theta)$ (see Figure 2.3). Suppose also that $P \notin L_3$ (the case $P \in L_3$ is easier). The problem then is to learn how to compose $\mathrm{Rot}(P, 2\theta)$ with $\rho_3 = \mathrm{Refl}(L)$.

In Figure 2.7, L is the axis of the third reflection ρ_3, and $Q = \rho_3(P)$. Draw the line M passing through the midpoint of PQ, such that the angle from M to L is θ; if we consider the rectangle $PAQB$ with PQ as a diagonal line, and sides PA and BQ parallel to M, it is easy to see that $\mathrm{Refl}(L) \circ \mathrm{Rot}(P, 2\theta) = \mathrm{Glide}(M, \mathbf{v})$ is the glide with axis the line M and translation vector the median vector $\mathbf{v}$.   QED

2.8 Preview of transformation groups

As we have seen in this chapter, the composite of maps $g \circ f$ is a basic, simple and familiar idea having many useful applications. From an algebraic point of view, the composite of Euclidean motions defines a product

$$\mathrm{Eucl}(n) \times \mathrm{Eucl}(n) \to \mathrm{Eucl}(n)$$

on the set Eucl(n) of motions of $\mathbb{E}^n$, which is associative (see 2.4), has an identity element and inverses. In other words, motions form a *transformation group* of $\mathbb{E}^n$. This idea is taken up again in Chapter 6 when we are ready for serious applications.

Exercises

2.1   A standard result of linear algebra identifies an $m \times n$ matrix $A = (a_{ij})$ with a linear map $\alpha \colon \mathbb{R}^n \to \mathbb{R}^m$ (taking the standard basis of column vectors to the columns of A). If $B = (b_{jk})$ is an $l \times m$ matrix giving a linear map $\beta \colon \mathbb{R}^m \to \mathbb{R}^l$, verify that the product matrix BA corresponds to the composite $\beta \circ \alpha$.

2.2   The (nonzero) complex numbers can be viewed as a set of *similarities* of $\mathbb{E}^2$: regard $z = x + iy$ as the map $T_z \colon \mathbb{R}^2 \to \mathbb{R}^2$ given by the matrix $\left(\begin{smallmatrix} x & y \\ -y & x \end{smallmatrix} \right)$. Write $z = r \exp(i\theta)$ where $r = |z|$ and $\theta = \arg z$, and interpret the map T_z geometrically. Prove that T_z is a similarity in the sense that there exists λ for which $d(T(x), T(y)) = \lambda d(x, y)$. Show how to obtain multiplication of complex numbers as composition of similarities.

2.3   In the notation of 2.4, prove that $h \circ g \circ f = (h \circ g) \circ f$. Prove that for 4 consecutive maps f, g, h, k, we have

$$(k \circ h) \circ (g \circ f) = k \circ ((h \circ g) \circ f).$$

Generalise the statement to any number of maps and any bracketing. Please be sure to dispose of your solution in the paper recycling bin.

2.4   In the notation of 2.4, find the conditions for the domain and range of f, g so that the commutative law

$$g \circ f \overset{?}{=} f \circ g$$

makes sense as a question. Show that the commutative law holds for the set of translations in $\mathbb{E}^n$, as well as the set of rotations of $\mathbb{E}^2$ about a fixed point. Show that it does not hold for the set of all motions of Euclidean space $\mathbb{E}^n$.

2.5   Verify by calculation that the usual definition of matrix multiplication $AB = (c_{ik} = \sum_j a_{ij} b_{jk})$ is associative. Use Exercise 2.1 and the associativity of maps to show that you do not need to do the calculation.

By 2.2, affine linear maps $T_{A,\mathbf{b}} \colon \mathbb{R}^n \to \mathbb{R}^n$ compose according to the rule $T_{A_2, \mathbf{b}_2} \circ T_{A_1, \mathbf{b}_1} = T_{A_2 A_1, A_2 \mathbf{b}_1 + \mathbf{b}_2}$; verify that this formula defines an associative multiplication rule.

Exercises in composing motions of $\mathbb{E}^2$.

2.6   The *half-turn* about P is the rotation through $180°$. Prove the following.
  (a)   The composite of 2 half-turns is a translation.
  (b)   Every translation is a composite of 2 half-turns.
  (c)   The composite of 3 half-turns is a half-turn.
  (d)   If L is a line and P a point then

$$\text{Refl}(L) \text{ and } \text{Halfturn}(P) \text{ commute} \iff P \in L.$$

2.7   Prove that every opposite motion of $\mathbb{E}^2$ is the composite of a half-turn and a reflection.

2.8   Give a geometric treatment of the composition of a rotation with a glide, to get another glide or reflection. When is Glide$(L, v) \circ \mathrm{Rot}\,\theta$ a reflection? [Hint: draw a diagram similar to Figure 2.7.]

2.9   Show that any composite $T_1 \circ T_2$ with either T_1 or T_2 a reflection or glide can be understood by drawing a diagram like Figure 2.7. [Hint: to view $g = \mathrm{Glide}(L, \mathbf{v})$ and its effect on a point $P \notin L$, draw a rectangle with the line $PT(P)$ as a diagonal and $\mathbf{v}$ as a median. The best way to see $g_1 \circ g_2$ is to draw two such rectangles with a common diagonal and the vectors $\mathbf{v}_1, \mathbf{v}_2$ as respective medians. For glide composed with rotation or translation, you guess that the answer is $g_1 \circ t = g_2$, which you can rewrite as $T = g_1^{-1} \circ g_2$ and treat similarly.]

2.10 (Harder) Use Claim 2.6 to study motions of $\mathbb{E}^3$ fixing a point O, and compare with the conclusion of Theorem 1.11. [Hint: a composite of 2 reflections in planes Π_1, Π_2 through O is a rotation about a line through O. For 3 reflections, you need to prove that $\mathrm{Refl}(\Pi) \circ \mathrm{Rot}(L, \theta)$ is a rotary reflection, or in other words, to find a plane which is rotated into itself by the composite.]

2.11 (Harder) Give a proof of Theorem 1.15 using Theorem 2.6. In other words, study the possibilities for the composite of ≤ 4 reflections of $\mathbb{E}^3$, and show that they lead to the 6 cases listed in Theorem 1.15. [Hint: see Rees [19].]

2.12 You can move a heavy piece of furniture (e.g. a bedroom wardrobe) by lifting the front and rotating it about the two back corners. Convince yourself that you can 'walk' your wardrobe anywhere in the Euclidean plane. (Ignore doors and stairs.)

  Let $P, Q \in \mathbb{E}^2$ be two distinct points. Prove that every direct motion of $\mathbb{E}^2$ is a composite of sufficiently many rotations about P and Q.

  [Hint: what kind of answer is required? First show that it is enough to prove that you can carry out any translation and any rotation about P. For the translations, think how you shift your wardrobe – easy does it!]

3 Spherical and hyperbolic non-Euclidean geometry

Together with plane Euclidean geometry, spherical and hyperbolic geometry are 2-dimensional geometries with the following properties:

(1) distance, lines and angles are defined and invariant under motions;
(2) the motions act transitively on points and directions at a point;
(3) locally, incidence properties are as in plane Euclidean geometry.

In more detail, (2) means that if P, P' are points, and λ, λ' directions at these points, then there exists a motion T taking P to P' and λ to λ'; in other words, the geometry is homogeneous (the same at every point) and isotropic (the same in every direction). (3) means that in sufficiently small open sets, a line is uniquely specified by a point and a direction, or by two points P, Q, and two lines l_i meet in at most one point (see Figure 3.0).

However, the geometries also differ in several respects:

(1) the global incidence properties of lines, that is, the existence of parallel and non-intersecting lines;
(2) intrinsic curvature properties: the perimeter of a circle, and the sum of angles in a triangle;
(3) the possibility of defining a unit of length intrinsic to the geometry.

Euclidean geometry in the plane was described in detail in Chapter 1. Although certainly not the same thing as plane geometry, spherical geometry is still very intuitive, because every definition and statement can be readily visualised on the very concrete model $S^2 \subset \mathbb{R}^3$, which you can hold in your hand or kick around a playing field. I discuss spherical lines (great circles), distances, angles and triangles, the classification of motions in terms of rotations and reflections, frames of reference and angular excess.

In contrast, plane hyperbolic geometry originally arose in axiomatic geometry (compare 9.1.2); the coordinate model I treat in this chapter is not immediately familiar, and was discovered many decades after axiomatic hyperbolic geometry. Although my model of hyperbolic geometry is not intuitive, essentially every step in my treatment is parallel to spherical geometry. Once you are sure you know what you are

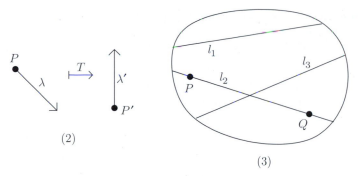

Figure 3.0   Plane-like geometry.

doing, you can just replace $x^2 + y^2 = 1$ by $-t^2 + x^2 = -1$, and the trig functions sin and cos by the hyperbolic trig functions sinh and cosh, and everything extends more or less word-for-word. This is the essential content of the prophetic suggestion by J. H. Lambert (1728–1777) that non-Euclidean geometry 'should be related to the geometry on a sphere of radius i $= \sqrt{-1}$' (see Coxeter [5], p. 299).

In Chapter 1 on Euclidean geometry, I discussed n-dimensional Euclidean space $\mathbb{E}^n$ along with the more familiar planar version. There is no logical reason to discontinue this practice, but for ease of digestion as well as notation, all definitions in this chapter are given in two dimensions. You will benefit immensely by generalising the definitions and, in some cases, the theorems to the higher dimensional setup; you are explicitly encouraged to do so in Exercise 3.10. Higher dimensional spheres appear in later chapters (see for example 7.4.2 and 8.5); unfortunately there is no space in the book for a detailed treatment of higher dimensional hyperbolic space and a discussion of its significance.

3.1 Basic definitions of spherical geometry

The sphere $S^2 \subset \mathbb{R}^3$ of radius r centred at the origin O is defined by the equation $x^2 + y^2 + z^2 = r^2$. I will often refer to points $P \in S^2$ via their position vector $\overrightarrow{OP} = \mathbf{p}$. A *spherical line* or *great circle* in S^2 is the intersection of S^2 with a plane $\Pi = \mathbb{R}^2$ through the origin; thus it is a circle in Π centred at O and with the same radius r as S^2. Two points $P, Q \in S^2$ are *antipodal* if their position vectors $\mathbf{p}, \mathbf{q}$ satisfy $\mathbf{p} = -\mathbf{q}$. Through any two distinct points $P, Q \in S^2$ which are not antipodal, there is a unique great circle or spherical line $L = PQ$. The *(spherical) distance* $d(P, Q)$ between points $P, Q \in S^2$ is the distance measured along the shorter arc of a great circle through P and Q; that is, it is radius r times $\angle POQ$, the angle at O between OP and OQ, where the angle is always interpreted as the absolute value in the range $[0, \pi]$. For ease of notation, I usually fix the radius $r = 1$ from now on.

Remarks

(1)   If you go back to the chapter on Euclidean geometry and compare the treatment of 1.1–1.3 to the one given here, you may notice that I have been a bit sloppy here. To

be consistent, I should have defined 'model' S^2 to be the sphere $\{x^2 + y^2 + z^2 = r^2\}$ in $\mathbb{R}^3$ with its inherited spherical distance, and 'abstract' S^2 to be a metric space isometric to 'model' S^2 but without a fixed choice of identification. Spelling this out explicitly leads to rather clumsy notation, but implicitly I am still following this procedure; in particular, I reserve the right to choose different coordinates on my 'abstract' metric S^2 if so needed. This remark applies equally well to the treatment of hyperbolic geometry in 3.9 below.

(2)   The sphere S^2 is defined as the subset $\{x^2 + y^2 + z^2 = 1\}$ of $\mathbb{R}^3$. On the northern hemisphere $\{z \geq 0\}$ I can rewrite this as $z = \sqrt{1 - x^2 - y^2}$. This gives a fairly good coordinate representation of S^2 near the north pole, but a fairly bad one in moderate or tropical regions. What is wrong with it? Well, if the model is the whole of $\mathbb{R}^2$, it is much too big; if we take only the disc $D^2 : x^2 + y^2 \leq 1$, crossing the equator in S^2 corresponds to falling off the edge of the world in the model. Furthermore, distances, angles, areas, curvature are all screwed up.

It is a basic problem in cartography to map regions of the surface of the Earth onto a plane. However, the map based on $z = \sqrt{1 - x^2 - y^2}$ is one of the most primitive and useless ways to do this. Over the course of time, several much better ways have been invented; see the references in the introduction of Chapter 9 for a starting point on this.

(3)   The distance $d(P, Q)$ is defined as (radius times) the angle of the PQ arc, $\alpha = \angle POQ$. It is useful to know how to translate between this *angle* and the *coordinates* of P, Q. In vector notation, the dot product of unit vectors equals the cosine of the angle between them: that is, if P, Q have position vectors $\mathbf{p}, \mathbf{q}$ then $\alpha = \angle POQ$ is given by

$$\mathbf{p} \cdot \mathbf{q} = \cos \alpha, \quad \text{that is,} \quad d(P, Q) = \alpha = \arccos(\mathbf{p} \cdot \mathbf{q}). \tag{1}$$

(I have set $r = 1$, so that $\mathbf{p}$ and $\mathbf{q}$ are unit vectors.) Recall that $\arccos = \cos^{-1}$ is the inverse function of cos, so that $\alpha = \arccos x$ is defined by the property $x = \cos \alpha$; similarly for arcsin. Here I choose α in the range $[0, \pi]$.

Given P and Q, I can choose coordinates so that $P = (0, 0, 1)$ and OPQ is the (x, z)-plane $\{y = 0\}$; then $Q = (\sin \alpha, 0, \cos \alpha)$. This is a parametrisation of the great circle, with parameter α. Points with $x < 0$ can also be included, by allowing $\alpha < 0$ to run through the range $[-\pi, \pi]$, but then $d(P, Q) = |\alpha|$.

In fact $(\sin \alpha, 0, \cos \alpha)$ is a parametrisation *by arc length*: if you think of (part of) the sphere S^2 as the graph of $z = \sqrt{1 - x^2 - y^2}$ as in (2), then $d(P, Q) = \int_P^Q ds$ where the infinitesimal arc length ds is determined by $ds^2 = dx^2 + dy^2 + dz^2$. Thus the length of arc PQ is

$$\int_0^{\sin \alpha} \frac{dx}{\sqrt{1 - x^2}} = \arcsin(\sin \alpha) = \alpha.$$

Geometers like to distinguish the *intrinsic* geometric properties of S^2 from those related to the embedding $S^2 \subset \mathbb{R}^3$. It is important in this context to notice that the natural distance in spherical geometry is the intrinsic distance, that is, the length of a certain curve traced in the surface S^2, as opposed to the distance in the ambient Euclidean space; you go from London to Singapore by plane, not by tunnel.

3.2 Spherical triangles and trig

The convention $r = 1$ is still in force. A *spherical triangle* $\triangle PQR$ consists of 3 vertexes P, Q, R and 3 arcs of great circle PQ, PR, QR joining them. These do not have to be the shorter arcs; P, Q are allowed to be antipodal, and then you have to specify one of the great circles to be the arc PQ.

The *spherical angle a* at P between the two lines PQ and PR is equal to the dihedral angle between the two planes OPQ, OPR in $\mathbb{R}^3$, in other words it is the angle between two lines cut out by the two planes in an auxiliary plane orthogonal to OP. You can take this as a definition if you like, and then you do not have to worry about how the angle between two curves is defined. More precisely, the tangent plane to S^2 at P is the 2-plane $T_P S^2$ defined by $z = 1$, and the tangent vectors to the two curves PQ and PR are the two lines in $T_P S^2$ cut out by these two planes. They are orthogonal to the axis OP, so the angle between the two curves equals the dihedral angle a between the two planes.

Proposition (Main formula of spherical trig)   *The side QR of the triangle is determined by the other two sides PQ and PR and the dihedral angle a. More precisely, write*

$$\alpha = \angle QOR = d(Q, R), \quad \beta = \angle POQ = d(P, Q), \quad \gamma = \angle POR = d(P, R).$$

(Recall that I have fixed the radius $r = 1$.) Then

$$\cos \alpha = \cos \beta \cos \gamma + \sin \beta \sin \gamma \cos a. \tag{2}$$

Proof   Although the statement looks complicated, the proof is easy 3-dimensional coordinate geometry. In Figure 3.2, let Q' and R' be the points on great circles at distance $\pi/2$ from P, so that $\overrightarrow{OQ'}$ is orthogonal to OP. Choose coordinates (x, y, z) so that $P = (0, 0, 1)$ (the north pole), and the equator is given by $z = 0$. Then Q' is a point on the equator, so I can choose $Q' = (1, 0, 0)$, and $R' = (\cos a, \sin a, 0)$. This determines the coordinates of all the points in the figure; by definition of β, γ, the following relations hold between the position vectors:

$$\mathbf{q} = \cos \beta \mathbf{p} + \sin \beta \mathbf{q}' = (\sin \beta, 0, \cos \beta),$$
$$\mathbf{r} = \cos \gamma \mathbf{p} + \sin \gamma \mathbf{r}' = (\sin \gamma \cos a, \sin \gamma \sin a, \cos \gamma).$$

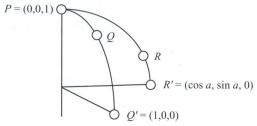

$P = (0,0,1)$

Q

R

$R' = (\cos a, \sin a, 0)$

$Q' = (1,0,0)$

Figure 3.2   Spherical trig.

Now α is the angle between the two unit vectors $\mathbf{q}$ and $\mathbf{r}$, so

$$\cos \alpha = \mathbf{q} \cdot \mathbf{r} = \cos \beta \cos \gamma + \sin \beta \sin \gamma \cos a. \quad \text{QED}$$

3.3   The spherical triangle inequality

Corollary (Triangle inequality)   *In any triangle $\triangle PQR$ whose sides are shorter arcs given by $\alpha, \beta, \gamma \leq \pi$ as above,*

$$\alpha \leq \beta + \gamma,$$

with equality if and only if PQR are collinear with P on the shorter arc QR.

Proof   This follows at once from the main formula (2) and calm reflection on the range of values for the angles α, β, γ and a. Notice that $\alpha, \beta, \gamma \in [0, \pi]$ essentially by convention: in defining distance I always take $\angle POQ$ to be the angle in the shorter arc. If β or $\gamma = 0$ or π, it is easy to read off the conclusion, so that I can assume that $\alpha, \beta, \gamma \in (0, \pi)$. On the other hand, in Figure 3.2, it is clear I want to have $a \in [0, 2\pi)$. Now compare (2) with the standard trig formula

$$\cos(\beta + \gamma) = \cos \beta \cos \gamma - \sin \beta \sin \gamma.$$

We know that $\sin \beta, \sin \gamma \in (0, 1]$; thus $\cos \alpha \geq \cos(\beta + \gamma)$, with equality if and only if $\cos a = -1$. Now $\cos \alpha$ is a strictly decreasing function in the range $[0, \pi]$, so that $\cos \alpha \geq \cos(\beta + \gamma)$ gives $\alpha \leq \beta + \gamma$. Equality holds only under the aforestated condition $\cos a = -1$, that is, if the short arcs PQ and PR are opposite when viewed from P.   QED

It is trivial that $d(P, Q)$ is symmetric, nonnegative, and positive unless $P = Q$, so that Corollary 3.3 proves that S^2 with the spherical distance is a metric space (see Appendix A).

3.4   Spherical motions

A *spherical motion* or *isometry* is of course just a map $T : S^2 \to S^2$ preserving spherical distance.

Theorem

(1)   A motion $T : S^2 \to S^2$ takes pairs of antipodal points to pairs of antipodal points, and spherical lines (great circles) to spherical lines.

(2)   Any motion is given in coordinates by $\mathbf{x} \mapsto A\mathbf{x}$, where A is a 3×3 orthogonal matrix.

Proof   Two points of the sphere are antipodal if and only if they are a maximum distance apart (at distance πr, half a world away), so the first sentence is clear. The rest of the proof is very similar to the Euclidean proof in Chapter 1. For (1), exactly as in Corollary 1.7, the arcs of spherical lines $[P, Q]$ are determined purely by the metric: three points P, Q, R are collinear (that is, on a spherical line or great circle) if and only if

$$d(P, Q) + d(Q, R) + d(R, P) = 2\pi r \quad \text{or} \quad \pm d(P, R) \pm d(R, Q) \pm d(P, Q) = 0.$$

Here the first equality is the statement that P, Q, R are on a great circle and not in any shorter great arc, and the second is the equality case of Corollary 3.3 for some permutation of P, Q, R. A spherical motion T preserves these equalities, so takes a spherical line L to a spherical line $L' = T(L)$.

For (2), note first that because $T : S^2 \to S^2$ takes antipodal points to antipodal points, it extends in a unique way to a map $T : \mathbb{R}^3 \to \mathbb{R}^3$ by radial extension. I claim that T is linear. For this, it is enough to see that T is linear when restricted to any plane Π through the origin.

Suppose $L = \Pi \cap S^2$ and $T(L) = L' = \Pi' \cap S^2$. A spherical line $L = \Pi \cap S^2$ is parametrised by arc length: a variable point of L is $\cos\theta\mathbf{f}_1 + \sin\theta\mathbf{f}_2$, where $\mathbf{f}_1, \mathbf{f}_2, \mathbf{f}_3$ is an orthogonal basis of $\mathbb{R}^3$ with $\mathbf{f}_1, \mathbf{f}_2 \in L$, and θ equals the arc length along L. Since T preserves distance, it preserves arc length along a spherical line, so that its restriction $T_L : L \to L'$ is given by

$$T(\cos\theta\mathbf{f}_1 + \sin\theta\mathbf{f}_2) = \cos\theta\mathbf{f}'_1 + \sin\theta\mathbf{f}'_2.$$

Here $\mathbf{f}'_1, \mathbf{f}'_2, \mathbf{f}'_3$ is a new orthogonal frame, with $\mathbf{f}'_1 = T(\mathbf{f}_1)$ and $\mathbf{f}'_2 = T(\mathbf{f}_2) \in L'$. Stated differently, $T(\lambda\mathbf{f}_1 + \mu\mathbf{f}_2) = \lambda\mathbf{f}'_1 + \mu\mathbf{f}'_2$, so T is linear.   QED

3.5   Properties of S^2 like $\mathbb{E}^2$

The following statements are either obvious, or can be done as easy exercises. Use them to refresh your memory of the case of $\mathbb{E}^2$, or as a warm-up for the case of the hyperbolic plane $\mathcal{H}^2$. The spherical statements are if anything a little simpler: for example, the distinction between translation and rotation disappears, and the classification of motions comes directly from the normal form of Theorem 1.11.

(1)   The sphere S^2 is a metric geometry with a distance function $d(P, Q)$, and motions given by 3×3 orthogonal matrixes.

(2)   The motions act transitively on S^2 and on spherical lines through a given point $P \in S^2$.

(3)   Every motion of S^2 is either a rotation $\mathrm{Rot}(P, \theta)$, or a reflection $\mathrm{Refl}(L)$ in a line (= great circle) or a glide $\mathrm{Glide}(L, \theta)$ (the restriction of a Euclidean rotary reflection).

(4)   Given two pairs of points P, Q and P', Q', there exist exactly two motions g of S^2 such that $g(P) = g(P')$, $g(Q) = g(Q')$, of which one is a rotation and the other a reflection or glide.

(5)   Motions come in two kinds, direct and opposite. Every direct motion is the identity or a composite of 2 reflections; every opposite motion is a reflection or a composite of 3 reflections.

(6)   The spherical distance $d(P, Q)$ between two points $P, Q \in S^2$ is the length of the shortest curve C in S^2 joining P and Q.

3.6   Properties of S^2 unlike $\mathbb{E}^2$

(1)   Incidence of lines. Any two spherical lines intersect in a pair of antipodal points. (Proof: if $L_1 = \Pi_1 \cap S^2$ and $L_2 = \Pi_2 \cap S^2$, consider the Euclidean line $\Pi_1 \cap \Pi_2$ in $\mathbb{R}^3$.) Therefore *spherical geometry has no parallel lines.*

(2)   Intrinsic distance. If you live on S^2, it makes sense to take the circumference of S^2 (or the length of any great circle) as a unit of distance; recall that the kilometre, adopted during the French revolution, was defined by setting the circumference of our own parochial sphere to be 40 000 km. Another aspect of the same phenomenon is that distances are bounded: $d(P, Q) \le \pi r \ (=: 20\,000 \text{ km})$.

(3)   Spherical frames. If you try to define a spherical frame of reference by analogy with the Euclidean notion, you get involved with the intrinsic distance. For example, if your unit of measurement is very big compared to the radius of the sphere, you will end up with your unit vector $P_0 Q_0$ wrapping the sphere several times. Taking a small unit of measurement, you can define a spherical frame $P_0 P_1 P_2$ and prove the analogue of Corollary 1.13 (a motion takes any frame into any other, and is uniquely determined by what it does to a frame) as an easy exercise. But there is an even better solution, which actively exploits the intrinsic distance: I can take the length $P_0 P_1$ to be 1/4 of the circumference, and get a spherical frame which coincides with an orthonormal frame of the ambient $\mathbb{R}^3$, so that the result about motions and frames is contained in Corollary 1.13.

(4)   Intrinsic curvature. To say that the sphere $S^2 \subset \mathbb{R}^3$ is curved, you could calculate the radius of curvature of lines relative to the ambient space $\mathbb{R}^3$. However, the geometry of S^2 also displays intrinsic curvature, as you can see in several ways. In $\mathbb{E}^2$ the perimeter of a Euclidean circle of radius ρ is $2\pi\rho$. By contrast, a spherical circle of radius ρ has perimeter $2\pi \sin \rho$, as discussed in Exercise 3.1.

(5)   Sum of angles in a triangle. Let S^2 be the sphere of radius $r = 1$, and $\triangle PQR$ a spherical triangle. Then

$$\angle P + \angle Q + \angle R = \pi + \text{area} \triangle PQR.$$

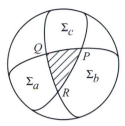

Figure 3.6 Overlapping segments of S^2.

Thus the sum of angles in a spherical triangle *never* equals $180°$. For very small triangles, you can view the discrepancy as a reflection of intrinsic curvature as in the preceding point.

Proof I prove the last point, because it is not obvious at first sight, and because the proof is very elegant. It is a 'Venn diagram' argument on the partition of S^2 obtained by slicing it up along the great circles which are the sides of $\triangle PQR$. Write Σ_a for the part of S^2 contained between the two planes OPQ and OPR (that is, the union of the two opposite segments) with a the dihedral angle between these planes, and similarly for Σ_b and Σ_c. Then by circular symmetry, clearly

$$\text{area } \Sigma_a = \frac{2a}{2\pi} \text{ area } S^2. \tag{3}$$

Now I claim that Σ_a, Σ_b, Σ_c cover S^2 and overlap exactly in $\triangle PQR$ and its antipodal triangle $\triangle P'Q'R'$ (see Figure 3.6).

Summing (3) for Σ_a, Σ_b and Σ_c gives

$$\text{area } S^2 + 4 \text{ area } \triangle = \text{area } \Sigma_a + \text{area } \Sigma_b + \text{area } \Sigma_c = (2a + 2b + 2c)\frac{\text{area } S^2}{2\pi}$$

(points in $\triangle$ and its antipodal triangle are covered 3 times, while the rest of S^2 is covered once). Therefore $a + b + c - \pi = (4\pi/\text{area } S^2) \text{ area } \triangle = \text{area } \triangle$.   QED

3.7 Preview of hyperbolic geometry

The remainder of this chapter introduces a coordinate model for hyperbolic geometry which is entirely parallel to spherical geometry. First, I review the ingredients of spherical geometry in one dimension.

(1) $\mathbb{R}^2$ with coordinates x, y and the ordinary Euclidean norm $x^2 + y^2$.
(2) The functions $\cos \theta = \frac{e^{i\theta} + e^{-i\theta}}{2}$ and $\sin \theta = \frac{e^{i\theta} - e^{-i\theta}}{2i}$, which satisfy the relation $\cos^2 + \sin^2 = 1$, and $\frac{d}{d\theta} \sin \theta = \cos \theta$, $\frac{d}{d\theta} \cos \theta = -\sin \theta$.
(3) The circle S^1 defined by $x^2 + y^2 = 1$ is parametrised by $x = \sin \theta$, $y = \cos \theta$, and the arc length is $\sqrt{dx^2 + dy^2} = d\theta$, so that θ is the arc length parameter for S^1.

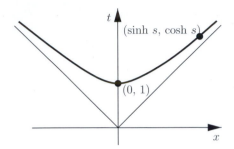

Figure 3.7   The hyperbola $t^2 = 1 + x^2$ and $t > 0$.

(4)   Symmetries are the set O(2) of rotation and reflection matrixes

$$\begin{pmatrix} \cos\theta & -\sin\theta \\ \sin\theta & \cos\theta \end{pmatrix} \quad \text{and} \quad \begin{pmatrix} \cos\theta & \sin\theta \\ \sin\theta & -\cos\theta \end{pmatrix}.$$

Now the ingredients of hyperbolic geometry in one dimension.

(1) $\mathbb{R}^2$ with coordinates t, x and the Lorentz pseudometric $-t^2 + x^2$. Here I choose a 'time-like' coordinate t and a 'space-like' coordinate x. A vector is *space-like* if it has positive squared length (for example $(0, x)$) and *time-like* if it has negative square (for example, $(t, 0)$ has squared length $-t^2$).

   The Lorentz space $\mathbb{R}^2$ is the ambient space for the hyperbola $\mathcal{H}^1$ defined by $t^2 = 1 + x^2$ and $t > 0$ (see Figure 3.7). The tangent space to $\mathcal{H}^1$ at any point $P_0 = (t_0, x_0) \in \mathcal{H}^1$ is the line $t = (x_0/t_0)x$, which is space-like, because $t_0 > |x_0|$. Therefore although the Lorentz pseudometric $-t^2 + x^2$ is not positive definite, the geometry of $\mathcal{H}^1$ itself contains only space-like directions.

(2)   The functions $\cosh s = \frac{e^s + e^{-s}}{2}$ and $\sinh s = \frac{e^s - e^{-s}}{2}$, which satisfy the relation $\cosh^2 - \sinh^2 = 1$, and $\frac{d}{ds}\sinh s = \cosh s$, $\frac{d}{ds}\cosh s = \sinh s$. It is useful to notice that sinh is a one-to-one map from the whole of $\mathbb{R}^1$ to the whole of $\mathbb{R}^1$.

(3)   The hyperbola $\mathcal{H}^1$ defined by $t^2 = 1 + x^2$ is parametrised by $x = \sinh s, t = \cosh s$, and the arc length in the Lorentz pseudometric is $\sqrt{-dt^2 + dx^2} = ds$, so that s is the arc length parameter for $\mathcal{H}^1$.

(4)   Symmetries are the set $O^+(1, 1)$ of Lorentz translation and reflection matrixes

$$\begin{pmatrix} \cosh s & \sinh s \\ \sinh s & \cosh s \end{pmatrix} \quad \text{and} \quad \begin{pmatrix} \cosh s & -\sinh s \\ \sinh s & -\cosh s \end{pmatrix}.$$

3.8   Hyperbolic space

Consider $\mathbb{R}^3$ with the Lorentz quadratic form $q_L(\mathbf{v}) = -t^2 + x^2 + y^2$ (compare B.2). The cone $\{q_L(\mathbf{v}) < 0\}$ breaks up into two subsets

$$\{t > +\sqrt{x^2 + y^2}\} \cup \{t < -\sqrt{x^2 + y^2}\}.$$

I fix the positive choice $t > 0$ throughout.

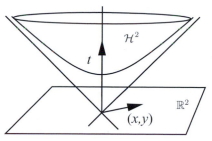

Figure 3.8 Hyperbolic space $\mathcal{H}^2$.

Hyperbolic space $\mathcal{H}^2 \subset \mathbb{R}^3$ is the upper sheet of the hyperboloid of two sheets given by $q_L(\mathbf{v}) = -1$:

$$\mathcal{H}^2 = \{(t, x, y) \mid -t^2 + x^2 + y^2 = -1 \text{ and } t > 0\}.$$

In other words, $t = \sqrt{1 + x^2 + y^2}$ (see Figure 3.8). This is the analogue of the sphere S^2 of radius 1, which is parametrised (in the northern hemisphere) by $z = \sqrt{1 - x^2 - y^2}$. If you want the analogue of the sphere of radius r, just take the hyperboloid $q_L(\mathbf{v}) = -r^2$. The coordinate t on $\mathbb{R}^3$ is 'time-like' and the coordinates x, y are 'space-like' (compare 3.7).

A *line* L of hyperbolic geometry is the hyperbola $\mathcal{H}^1$ obtained as the intersection of $\mathcal{H}^2$ with a 2-dimensional vector subspace $\Pi \subset \mathbb{R}^3$ which is a *Lorentz plane*, in the sense that it contains time-like vectors, so that $L = \Pi \cap \mathcal{H}^2 \neq \emptyset$; the restriction of q_L to Π has signature $(-1, +1)$. It is obvious that there is a unique line PQ through any two distinct points $P, Q \in \mathcal{H}^2$, since the 2-dimensional vector subspace Π through P, Q in $\mathbb{R}^3$ is unique. The analogy with the lines of S^2 is clear, and I could reasonably call the lines of L *great hyperbolas*.

3.9 Hyperbolic distance

To define the hyperbolic distance function, I start with the formal analogue of formula (1) of Remark 2 in 3.1, replacing the Euclidean inner product with the Lorentz inner product $\cdot_L$ (see B.2). Thus let P and Q be points of $\mathcal{H}^2$ given by the vectors $\mathbf{v} = (t_1, x_1, y_1)$ and $\mathbf{w} = (t_2, x_2, y_2)$. I define the *hyperbolic distance* $d(P, Q)$ between two points by

$$-\mathbf{v} \cdot_L \mathbf{w} = \cosh d(P, Q), \quad \text{so that} \quad d(P, Q) = \operatorname{arccosh}(-\mathbf{v} \cdot_L \mathbf{w}); \quad (4)$$

in other words, $d(P, Q) = \operatorname{arccosh}(t_1 t_2 - x_1 x_2 - y_1 y_2)$.

Lemma *The Lorentz inner product satisfies*

$$-\mathbf{v} \cdot_L \mathbf{w} = t_1 t_2 - x_1 x_2 - y_1 y_2 \geq 1,$$

with equality only if $P = Q$. (See also Exercise 3.11.) Hence the distance $d(P, Q)$ is defined and positive unless $P = Q$.

Proof   This clearly follows from the stronger statement.

Claim   *Given two points* $P \neq Q \in \mathcal{H}^2$, *there is a Lorentz basis* $\mathbf{f}_0, \mathbf{f}_1, \mathbf{f}_2$ *of* $\mathbb{R}^3$ *giving rise to a new coordinate system in which* $P = (1, 0, 0)$ *and* $Q = (\cosh \alpha, \sinh \alpha, 0)$, *with* $\alpha = d(P, Q) > 0$.

This is simply Appendix B, Theorem B.3 (4), but I need one point of the proof, so I repeat it here. Set $\mathbf{f}_0 = \mathbf{v}$ the position vector of P; since $P \in \mathcal{H}^2$, this vector has Lorentz norm -1. The vector $\mathbf{w}' = \mathbf{w} + (\mathbf{w} \cdot_L \mathbf{f}_0)\mathbf{f}_0$, where $\mathbf{w}$ is the position vector of Q, is orthogonal to $\mathbf{f}_0$ with respect to $\cdot_L$ (just compute the product $\mathbf{w}' \cdot_L \mathbf{f}_0$)), and is nonzero because $P \neq Q$. Hence by Theorem B.3 (3), $q_L(\mathbf{w}') > 0$. So I can set $\mathbf{f}_1 = \mathbf{w}'/\sqrt{q_L(\mathbf{w}')}$, and

$$\mathbf{w} = c\mathbf{f}_0 + s\mathbf{f}_1, \quad \text{where } c = -\mathbf{v} \cdot_L \mathbf{w} \text{ and } s = \sqrt{q_L(\mathbf{w}')} > 0. \tag{5}$$

I find the remaining basis element by the usual method of making an orthonormal basis: choose $\mathbf{u} \in \mathbb{R}^3$ not in the span of $\mathbf{v}$ and $\mathbf{w}$, set $\mathbf{w}'' = \mathbf{u} + (\mathbf{u} \cdot_L \mathbf{f}_0)\mathbf{f}_0 - (\mathbf{u} \cdot_L \mathbf{f}_1)\mathbf{f}_1$ and finally $\mathbf{f}_2 = \mathbf{w}''/\sqrt{q_L(\mathbf{w}'')}$.

The Lorentz basis $\mathbf{f}_0, \mathbf{f}_1, \mathbf{f}_2$ defines a new coordinate system on the hyperbolic plane $\mathcal{H}^2$. In this coordinate system $P = (1, 0, 0)$ and $Q = (c, s, 0)$, the latter by the first equality in (5). As $Q \in \mathcal{H}^2$, $c > 0$ and its position vector has Lorentz norm -1, so $-c^2 + s^2 = -1$. By (5), $s > 0$ and hence $c > 1$. So $c = \cosh \alpha$, $s = \sinh \alpha$ for some $\alpha > 0$, and in this coordinate system it is easy to compute $d(P, Q) = \alpha$. Hence the distance function is meaningful and positive unless $P = Q$.   QED

Compare Remark 3.1 (2) for the spherical analogy; the purist may want to reread Remark 3.1 (1) at this point.

Remark   This proof illustrates the fact that in the treatment of hyperbolic geometry given here, the methods of linear and quadratic algebra are our main weapons of attack. The arguments are similar to their Euclidean and spherical analogues, the only difference being the issue of the extra sign in the Lorentz form, along with the additional care it needs.

The question of signs is important later: in (5), $s = \sinh \alpha > 0$ was part of the construction of the vector $\mathbf{f}_1$. Notice that $\cosh \alpha$ is a symmetric function and $\sinh \alpha$ is an antisymmetric function. This is good, because I am measuring distances from the base point $P = (1, 0, 0)$ in terms of $\cosh \alpha$, and using $\sinh \alpha$ to parametrise the hyperbola by arc length α.

3.10   Hyperbolic triangles and trig

This section is the analogue of 3.2. A *hyperbolic triangle* $\triangle PQR$ in $\mathcal{H}^2$ consists of 3 vertexes P, Q, R and 3 hyperbolic lines PQ, PR, QR joining them. Choose coordinates as in Lemma 3.9 so that $P = (1, 0, 0)$ and PQ is on the hyperbolic line $\{y = 0\}$; set $Q' = (0, 1, 0)$.

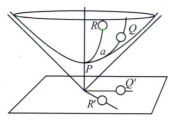

Figure 3.10   Hyperbolic trig.

The *hyperbolic angle a* at P between the two lines PQ and PR is defined to be the dihedral angle between the two planes OPQ, OPR (see Figure 3.10). The point is that this is a Euclidean angle, namely, the angle between two lines OQ' and OR' in the space-like plane $t = 0$; in other words, the line PR is in the plane OPR' spanned by P and $R' = (0, \cos a, \sin a)$.

Proposition (Main formula of hyperbolic trig)   *In a hyperbolic triangle $\triangle PQR$, the side QR is determined by the two sides PQ and PR and the dihedral angle a: if $\alpha = d(Q, R)$, $\beta = d(P, Q)$, $\gamma = d(P, R)$, then*

$$\cosh \alpha = \cosh \beta \cosh \gamma - \sinh \beta \sinh \gamma \cos a. \tag{6}$$

Proof   In the notation developed above, $P = (1, 0, 0)$, $Q = (\cosh \beta, \sinh \beta, 0)$ and

$$R = (\cosh \gamma, \sinh \gamma \cos a, \sinh \gamma \sin a);$$

here, as in (5), $\sinh \gamma > 0$ is part of the definition of the angle a. Thus calculating the Lorentz dot product of the two vectors representing Q and R gives

$$\cosh \alpha = \cosh \beta \cosh \gamma - \sinh \beta \sinh \gamma \cos a. \quad \text{QED}$$

Corollary (Triangle inequality) $d(Q, R) \le d(P, Q) + d(P, R)$, *with equality if and only if P is on the interval $[Q, R]$ (that is, the segment of line joining Q and R).*

Proof   This is exactly as before: compare (6) with the standard formula of hyperbolic trig:

$$\cosh(\beta + \gamma) = \cosh \beta \cosh \gamma + \sinh \beta \sinh \gamma.$$

Both $\sinh \beta$ and $\sinh \gamma$ are positive, so that $\cosh(\beta + \gamma) \ge \cosh \alpha$, with equality if and only $a = \pi$. Since $\cosh \alpha$ is an increasing function for $\alpha > 0$, it follows that $\beta + \gamma \ge \alpha$, with equality if and only if $P \in [Q, R]$.   QED

Remark   An important corollary of the triangle inequality, in complete analogy with Euclidean and spherical geometry, is the fact that the hyperbolic distance $d(P, Q)$

between two points $P, Q \in \mathcal{H}^2$ is the length of the shortest curve C in $\mathcal{H}^2$ joining P and Q, this shortest curve being the hyperbolic line segment $[P, Q]$. The proof, with the usual assumptions about the meaning of the statement, is word for word the same as in 1.4.

3.11 Hyperbolic motions

A *hyperbolic motion* $T : \mathcal{H}^2 \to \mathcal{H}^2$ is a map preserving hyperbolic distance. As before, my first aim is to get from this definition to a manageable description of T in terms of a suitable matrix. Read the homework on Lorentz matrixes in B.4–B.5, before you continue.

Theorem

1. *Every hyperbolic motion preserves hyperbolic lines.*
2. *Every hyperbolic motion $T : \mathcal{H}^2 \to \mathcal{H}^2$ is given in coordinates by $\mathbf{x} \mapsto A\mathbf{x}$, where*
    (a) A is a Lorentz matrix, that is

$$
{}^{t}A \begin{pmatrix} -1 & 0 & 0 \\ 0 & 1 & 0 \\ 0 & 0 & 1 \end{pmatrix} A = \begin{pmatrix} -1 & 0 & 0 \\ 0 & 1 & 0 \\ 0 & 0 & 1 \end{pmatrix}, \quad and
$$

(b) A preserves the two halves of the cone $\{q_L(\mathbf{v}) < 0\}$.

Proof   The proofs are almost the same as in the Euclidean and spherical cases (see 1.7 and Theorem 3.4 (2)). Since lines are determined by the distance function, a motion T takes a hyperbolic line to another hyperbolic line, proving (1). Since a hyperbolic line L is a hyperbolic arc in a Lorentz plane $\Pi = \mathbb{R}^2$ with arc length parametrisation $(\cosh s, \sinh s)$, it follows that T is linear when restricted to each Π, therefore linear on $\mathbb{R}^3$.

More formally, I can extend T from $\mathcal{H}^2$ to the upper half-cone by radial extension; write $\widetilde{T}$ for this extension. Give a Lorentz plane Π, choose a Lorentz basis $\mathbf{f}_0, \mathbf{f}_1$ so that L is parametrised as

$$
P_s = (\cosh s)\mathbf{f}_0 + (\sinh s)\mathbf{f}_1 \quad \text{for } s \in \mathbb{R};
$$

here the time-like vector $\mathbf{f}_0$ is the coordinate of a point $P_0 \in L$, and the space-like vector $\mathbf{f}_1$ is the tangent direction to L at P_0, with s the distance function along L. Then T takes L to the line L' parametrised as $P'_s = (\cosh s)\mathbf{f}'_0 + (\sinh s)\mathbf{f}'_1$, so that $\widetilde{T}$ is given by a linear map on Π. Since this holds for any line L, it follows that $\widetilde{T}$ is linear within the upper half-cone (that is,

$$
\widetilde{T}(\lambda\mathbf{u} + \mu\mathbf{v}) = \lambda\widetilde{T}(\mathbf{u}) + \mu\widetilde{T}(\mathbf{v})
$$

whenever $\mathbf{u}, \mathbf{v}$ and $\lambda\mathbf{u} + \mu\mathbf{v}$ are in the upper half-cone). Now, although $\widetilde{T}$ is only defined in the half-cone, the usual linear algebra argument shows that it is given by a matrix A (just choose a basis of the vector space $\mathbb{R}^3$ consisting of three vectors in the upper half-cone). Moreover, A must be Lorentz since $\widetilde{T}$ preserves the Lorentz form (compare B.4–B.5). QED

Remark   In proving Theorem 3.4, I extended T to $\mathbb{R}^3$ by radial extension, then used linearity on each plane Π, which holds because the distance function determines everything about motions in 1 dimension. In the hyperbolic case, the awkward point is that radial extension only gives $\widetilde{T}$ defined on the upper half-cone; my argument is that it is linear in the upper half-cone, and so given by a matrix.

A Lorentz matrix A preserves the two halves of the cone $\{q_L(\mathbf{v}) < 0\}$ if and only if its top left entry $a_{00} > 0$; such a matrix defines a *Lorentz transformation* of $\mathbb{R}^3$. The set $O^+(1, 2)$ of Lorentz transformations is entirely analogous to the set Eucl(2) of motions of the Euclidean plane. It is easy to state and prove the following assertions, all of which are analogues of the corresponding statements in plane Euclidean geometry (compare also 3.5).

1. The hyperbolic plane $\mathcal{H}^2$ is a metric geometry with a distance function $d(P, Q)$ and a set of motions $O^+(1, 2)$.
2. The motions act transitively on $\mathcal{H}^2$ and the set of lines through a given point $P \in \mathcal{H}^2$.
3. Every element of $O^+(1, 2)$ is either a rotation $\mathrm{Rot}(P, \theta)$, a Lorentz translation $\mathrm{Transl}(L, \alpha)$ along an axis L, a Lorentz reflection $\mathrm{Refl}(L)$ or a Lorentz glide. For example, if $L = \{y = 0\}$, the translation and glide are given by

$$\begin{pmatrix} \cosh s & \sinh s & 0 \\ \sinh s & \cosh s & 0 \\ 0 & 0 & 1 \end{pmatrix} \quad \text{and} \quad \begin{pmatrix} \cosh s & \sinh s & 0 \\ \sinh s & \cosh s & 0 \\ 0 & 0 & -1 \end{pmatrix}.$$

(Compare Exercise B.3.)

4. Given two pairs of points P, Q and P', Q', there exist exactly two motions $g \in O^+(1, 2)$ such that $g(P) = g(P'), g(Q) = g(Q')$, of which one is a rotation or Lorentz translation and the other a Lorentz reflection or glide.
5. $O^+(1, 2)$ has two types of elements, direct and indirect. Every direct motion is the identity or a composite of 2 reflections; every opposite motion is a reflection or a composite of 3 reflections.

3.12 Incidence of two lines in $\mathcal{H}^2$

In 3.6 (1) I showed that two lines (great circles) of S^2 meet in a pair of antipodal points, by taking $L_1 = \Pi_1 \cap S^2, L_2 = \Pi_2 \cap S^2$, then constructing the line $V = \Pi_1 \cap \Pi_2$ in the ambient $\mathbb{R}^3$, which of course meets S^2 in two points. Two familiar facts follow: (1) the orthogonal complement $V^\perp \subset \mathbb{R}^3$ is a plane cutting out a line $M = V^\perp \cap S^2$, the unique common perpendicular to L_1 and L_2; (2) L_1, L_2 generate a pencil of lines, that pass through the same intersection points and are perpendicular to M. If I choose coordinates so that V is the z-axis, the intersection points are the poles $(0, 0, \pm 1)$, M

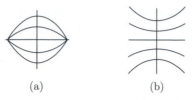

Figure 3.12   (a) Projection to the (x, y)-plane of the spherical lines $y = cz$. (b) Projection to the (x, y)-plane of the hyperbolic lines $y = ct$.

is the equatorial plane $z = 0$, and the family of lines containing L_1, L_2 is the pencil of meridians $(\sin\theta)x = (\cos\theta)y$ (Figure 3.12).

The same arguments apply to lines in $\mathcal{H}^2$, but the conclusions are different, since the ambient $\mathbb{R}^3$ is now Lorentz space: as before, let $L_1 = \Pi_1 \cap \mathcal{H}^2$, $L_2 = \Pi_2 \cap \mathcal{H}^2$, and consider the line $V = \langle\mathbf{v}\rangle = \Pi_1 \cap \Pi_2 \subset \mathbb{R}^3$. There are 3 cases.

(i) V is space-like: $q_L(\mathbf{v}) > 0$. Then L_1, L_2 are disjoint, since $V \cap \mathcal{H}^2 = \emptyset$. In this case, the orthogonal complement $V^\perp$ with respect to the Lorentz inner product $\cdot_L$ is a Lorentz plane (the restriction of q_L has signature $(-1, +1)$, so that it contains time-like vectors), and hence $M = V^\perp \cap \mathcal{H}^2$ is a line of $\mathcal{H}^2$, and is the unique common perpendicular to L_1, L_2. For example, if V is the x-axis, the lines L_1, L_2 are among the meridian lines $y = ct$, having the common perpendicular $M : (x = 0)$.

(ii) V is time-like: $q_L(\mathbf{v}) < 0$. Then L_1, L_2 intersect in $P = V \cap \mathcal{H}^2$. They do not have a common perpendicular, because the plane $V^\perp \subset \mathbb{R}^3$ is space-like, so does not meet $\mathcal{H}^2$. For example, if V is the t-axis, L_1, L_2 intersect at $P = (1, 0, 0)$ and the pencil of lines through P is $(\sin\theta)x = (\cos\theta)y$.

(iii) V is actually on the light cone: $q_L(\mathbf{v}) = 0$. Then L_1, L_2 are disjoint in $\mathcal{H}^2$, but are asymptotic, in the sense that they approach indefinitely at one end. For example, $V = \langle(1, 1, 0)\rangle$ is the common asymptotic direction of the lines $L_c : (y = c(t - x))$ with $|c| < 1$. The plane $V^\perp : (x = t)$ is tangent to the light cone along V, so does not correspond to a line in $\mathcal{H}^2$, and L_1, L_2 do not have a common perpendicular.

Definition   I say that L_1 and L_2 *diverge* in case (i). A simple calculation shows that, if L_1 and L_2 are parametrised by arc length as $P_1(s)$, $P_2(s)$ then $d(P_1(s), P_2(s))$ grows linearly in s as $s \gg 0$; for details, see Exercise 3.21.

Case (iii) is the limiting case that separates (i) and (ii): although L_1, L_2 are disjoint, they 'approach one another at infinity'. I say that L_1, L_2 are *ultraparallel*. To make this precise, it is useful to introduce the formal idea that each line $L = \Pi \cap \mathcal{H}^2$ of $\mathcal{H}^2$ has two 'ends', the two rays in which the plane Π intersects the null-cone $q(\mathbf{v}) = 0$, or the asymptotic lines of the hyperbola $L \subset \Pi$. One views an end as an 'ideal point' of L or 'point at infinity', not a point of $\mathcal{H}^2$, but rather an asymptotic direction. Case (iii) above, can be described by saying that L_1 and L_2 have a common end $V = \langle\mathbf{v}\rangle = \Pi_1 \cap \Pi_2$. By convention, ultraparallel lines L_1 and L_2 have angle 0 at this end. All the lines $L_c : y = c(t - x)$ are ultraparallel, with the ray $(1, 1, 0)$ as a common end. These lines all approach one another arbitrarily closely as they head out to infinity, as described in Exercise 3.20.

3.13 The hyperbolic plane is non-Euclidean

As discussed in the introduction to this chapter and at the end of 3.11, hyperbolic geometry shares many features with Euclidean and spherical geometry; the differences are also striking. The incidence properties of lines in $\mathcal{H}^2$ just established are qualitatively quite different from the Euclidean case. Two lines L_1 and L_2 of $\mathcal{H}^2$ have a common perpendicular M if and only if $V = \Pi_1 \cap \Pi_2$ is space-like, which is clearly an *open* condition: L_1 and L_2 remain disjoint even if we move them a little, for example, tilting one of them about a point. The parallel postulate thus fails, as I discuss below in more detail. The next section 3.14 treats the angular defect formula, expressing the sum of angles in a triangle in terms of its area; this sum is always $< \pi$. The hyperbolic non-Euclidean world also differs from the Euclidean in the existence of an intrinsic distance, by analogy with the spherical world (compare 3.6), and the negative curvature of hyperbolic space (compare Exercise 3.13 (c) and 9.4).

Euclid's parallel postulate states that given a line L of the planar geometry and a point P not on it, there is one and only one line M through P and disjoint from L. This holds in plane Euclidean geometry (and indeed in affine geometry, compare 4.3); in spherical geometry it is obviously false as there are no disjoint lines. What happens in $\mathcal{H}^2$? A plausible attempt to find a parallel line M through a point $P \notin L$ is to drop a perpendicular PQ onto L, then take M perpendicular to PQ; as we know from the above, this is indeed a line not meeting L, but not the only one.

Theorem   *Let L be a hyperbolic line and P a point not lying on L. Then there exists a unique perpendicular line PQ to L through P. Moreover,*

(1)   *if M is orthogonal to PQ in P, then the lines L and M diverge;*

(2)   *there exists an angle $\theta < \frac{\pi}{2}$ with the property that if L' is a line through P, then L' meets L if and only if the angle of L' and PQ at P is less than θ. (See Figure 3.13.)*

Remark   In axiomatic geometry, the logical self-consistency of this picture was the focal point of the 2000 year old controversy concerning Euclid's parallel postulate (compare 9.1.2). In the present coordinate construction of $\mathcal{H}^2$, there is nothing to dispute: everything follows at once from the case division of 3.12. Whether Euclidean or hyperbolic geometry or some other theory is a better approximate mathematical model for the real world in different applications is an entirely separate question, discussed in 9.4.

Proof   I give the coordinate proof. The line L corresponds to a Lorentz orthogonal decomposition $\mathbb{R}^3 = \Pi \oplus \Pi^\perp$ where $L = \Pi \cap \mathcal{H}^2$. The coordinate vector $\mathbf{p}$ of P can

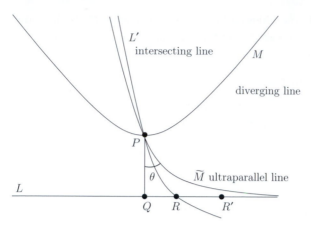

Figure 3.13   The failure of the parallel postulate in $\mathcal{H}^2$.

be written

$$\mathbf{p} = \mathbf{q} + \mathbf{v} \quad \text{with } \mathbf{q} \in \Pi \text{ and } \mathbf{v} \text{ in } \Pi^\perp;$$

here $\mathbf{v}$ is nonzero and space-like, and $\mathbf{q} \neq 0$ because $\mathbf{p}$ is time-like. Choosing Lorentz coordinates in $\mathbb{R}^3$ with $\mathbf{e}_0$ the unit time-like vector proportional to $\mathbf{q}$ and $\mathbf{f}_2$ proportional to $\mathbf{v}$ makes L into the line $y = 0$, $Q = (1, 0, 0)$ and $P = (t_0, 0, y_0)$ with $t_0 = \sqrt{1 + y_0^2}$. The perpendicular line PQ is $x = 0$, and the line M perpendicular to it at P is $y = \frac{y_0}{t_0} t$. The two planes of L and M intersect in the x-axis of $\mathbb{R}^3$, so L and M diverge.

Any line through $P = (t_0, 0, y_0)$ is given by $(\sin \varphi)x = (\cos \varphi)(y_0 t - t_0 y)$; in $\mathbb{R}^3$, this plane intersects $y = 0$ in the line $\langle (\tan \varphi, y_0, 0) \rangle$, which is time-like if and only if $|\tan \varphi| > y_0$. This proves the claim (together with the actual value $\theta = \operatorname{arccot} y_0$, compare Exercise 3.17).   QED

Discussion   A second 'proof' in more geometric terms is much closer to the historical context, if trickier to argue convincingly; please refer to Figure 3.13 during the argument. The existence and uniqueness of the orthogonal PQ can be proved by minimising the distance from P to L, as discussed in Exercise 3.15 (b); (1) follows from the case division in 3.12, and is proved again in Exercise 3.21.

For (2), note first that some lines L' through P certainly meet L. On the other hand, as (1) shows, there exists a line M through P that does not meet L. It is also easy to see that there cannot be a 'last' line L' through P which meets L: if $L' \cap L = R$ then there are points R' along L and further away from Q, and hence further lines PR' meeting L. From this, a least upper bound argument shows that there must be a 'first' line $\widetilde{M}$ (one on either side of PQ) which fails to meet L.

This proves almost all of (2); the only remaining point to clear up is the statement that the angle θ between PQ and the 'first' nonintersecting line $\widetilde{M}$ is less than $\pi/2$. However, the line M at angle exactly $\pi/2$ diverges from L by (1), whereas $\widetilde{M}$ is asymptotic to L; hence the angle θ must be less than $\pi/2$. Lines L' having angle less than θ at P with PQ are of type (i) and so intersect L; lines having angle greater than θ are of type (ii) and are disjoint from L.

Other models   There are several alternative models of non-Euclidean geometry in addition to the hyperbolic model in Lorentz space discussed here. Beltrami's model as the interior of an absolute conic in $\mathbb{P}^2_{\mathbb{R}}$ is treated in Rees [19]; it has the great advantage of making the incidence of lines completely transparent. An alternative is the Lobachevsky or Poincaré model as the upper half-space in the complex plane, which makes asymptotically converging ultraparallel lines easy to visualise, and which is important in other mathematical contexts; Exercises 3.23–25 lead you through the construction of this model.

3.14 Angular defect

The remainder of this chapter discusses two proofs of the famous angular defect formula of Gauss and Lobachevsky.

Theorem   *In a hyperbolic triangle $\triangle PQR$ with angles a, b, c,*

$$a + b + c = \pi - \text{area} \triangle PQR. \tag{7}$$

In addition to finite hyperbolic triangles $\triangle PQR$ with $P, Q, R \in \mathcal{H}^2$, I generalise the statement to allow *ideal triangles*, with one or more vertexes ideal points 'at infinity'. An *ideal triangle* has 3 sides which are lines of $\mathcal{H}^2$, and any 2 sides either intersect, or are ultraparallel in the sense of Definition 3.13, with every pair of sides intersecting in distinct (ideal) points. Remember that 2 lines meeting at an ideal point have angle 0 there.

3.14.1
The first
proof
I. First, an explicit integration calculates the area of the particular triangle $\triangle PQR$ of Figure 3.14a. The crucial point here is that the area of a triangle remains bounded, even though one of its vertexes goes off to infinity.

II. Next, area of polygons and sum of angles of polygons have the simple *additivity* property illustrated in Figure 3.14b: if you subdivide A as a union of two adjacent polygons $A = B \cup C$, then area $A = $ area $B + $ area C. The sum of angles also adds, except that you subtract π if two angles coalesce to form a straight line (because the common point is no longer viewed as a vertex).

3.14.2
An explicit
integral

Proposition   *Let $a \in (0, \pi/2)$ be a given angle. Consider $\triangle PQR$ in $\mathcal{H}^2$ bounded by the three lines $y = 0$, $y = (\tan a)x$ and $x = (\cos a)t$ (see Figure 3.14a). Then*

$$\text{area} \triangle PQR = \pi/2 - a = \pi - angle\ sum(\triangle PQR).$$

Proof   The triangle has two vertexes $P = (1, 0, 0)$ and $Q = \frac{1}{\sin a}(1, \cos a, 0)$ in $\mathcal{H}^2$ and one ideal vertex $R = (1, \cos a, \sin a)$. We know that $\angle RPQ = a$ for the same reason as in 3.2 and 3.10, because the angle in $\mathcal{H}^2$ is the dihedral angle in $\mathbb{R}^3$, which equals the angle in the plane $\{t = 0\}$. I have drawn Figure 3.14a with symmetry

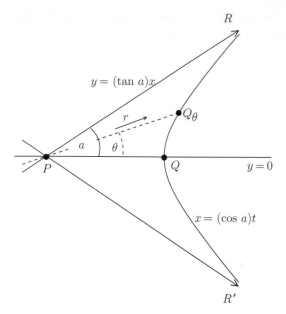

Figure 3.14a   The hyperbolic triangle $\triangle$ PQR with one ideal vertex.

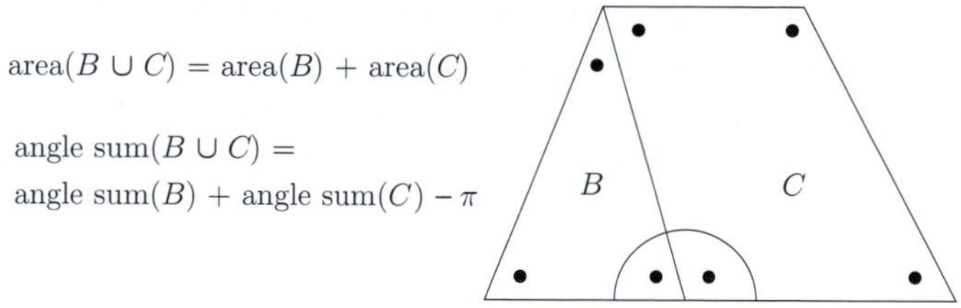

$$\text{area}(B \cup C) = \text{area}(B) + \text{area}(C)$$

$$\text{angle sum}(B \cup C) =$$
$$\text{angle sum}(B) + \text{angle sum}(C) - \pi$$

Figure 3.14b   Area and angle sums are 'additive'.

about the x-axis so that we see at once that $\angle PQR = \pi/2$. Finally, $\angle PRQ = 0$ by definition. Hence

$$\text{angle sum}(\triangle PQR) = \pi/2 + a$$

which proves the second equality.

To calculate the area, I write down an element of area, and integrate it as a double integral over the triangle $\triangle PQR$. It is convenient to work in polar coordinates

$$x = r \cos\theta, \quad y = r \sin\theta, \quad \text{so that } t = \sqrt{1 + r^2}.$$

In these coordinates, the element of area in $\mathcal{H}^2$ is $r \, dr \, d\theta / \sqrt{1 + r^2}$ (see Exercise 3.22 and compare also Exercise 3.8). It is easy to integrate this element of area as an indefinite integral, since

$$\frac{r \, dr}{\sqrt{1 + r^2}} d\theta = d\sqrt{1 + r^2} \, d\theta.$$

The more subtle point is to get an explicit expression for the domain of integration. Since the two sides out of P in Figure 3.14a are given by $y = 0, y = (\tan a)x$, the angle θ runs through the interval $[0, a]$. For fixed θ, the point $(\sqrt{1 + r^2}, r \cos \theta, r \sin \theta)$ runs through the line PQ_θ of Figure 3.14a. The condition to be under the hyperbola is $x \le (\cos a)t$, giving

$$r \cos \theta \le \sqrt{1 + r^2} \cos a \implies r^2 \le \frac{\cos^2 a}{\cos^2 \theta - \cos^2 a}.$$

Therefore

$$\text{area } \triangle PQR = \iint\limits_{\triangle PQR} \frac{r \, dr \, d\theta}{t} = \iint\limits_{\triangle PQR} d\sqrt{1 + r^2} \, d\theta$$

$$= \int_{\theta=0}^{a} \left[\sqrt{1 + r^2} \right]_{r^2=0}^{r^2 = \frac{\cos^2 a}{\cos^2 \theta - \cos^2 a}} d\theta$$

$$= \int_0^a \left(-1 + \sqrt{\frac{\cos^2 \theta}{\cos^2 \theta - \cos^2 a}} \right) d\theta.$$

Now I am in luck, and the integrand is an exact differential: indeed, consider $\varphi = \arcsin(\sin \theta / \sin a)$ as a function of θ. Then differentiating the defining relation $(\sin a)(\sin \varphi) = \sin \theta$ gives

$$\frac{d\varphi}{d\theta} = \frac{\cos \theta}{(\sin a)(\cos \varphi)} = \sqrt{\frac{\cos^2 \theta}{\cos^2 \theta - \cos^2 a}}.$$

It follows that the above integral evaluates to

$$\text{area } \triangle PQR = -a + \left[\arcsin \left(\frac{\sin \theta}{\sin a} \right) \right]_0^a = -a + \pi/2. \quad \text{QED}$$

3.14.3
Proof by
subdivision

The calculation of Proposition 3.14.2 implies at once the following result for ideal triangles with two or more ideal vertexes.

Lemma

(1)   *Let $\triangle PRR'$ be an ideal triangle of $\mathcal{H}^2$ with one vertex $P \in \mathcal{H}^2$ and two ideal vertexes; if $\angle P = a$ then*

$$\text{area } \triangle PRR' = \pi - a. \tag{8}$$

(2)   *Let $\triangle PQR$ be an ideal triangle of $\mathcal{H}^2$ with all three vertexes P, Q, R ideal points at infinity. Then*

$$\text{area } \triangle PQR = \pi. \tag{9}$$

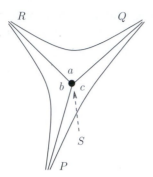

Figure 3.14c   The subdivision of $\triangle$ PQR.

Proof   (1) Drop a perpendicular PQ from P onto the opposite side RR'. By Claim 3.9, I can choose coordinates such that $P = (1, 0, 0)$ and PQ is the x-axis $y = 0$. This subdivides triangle $\triangle PRR'$ symmetrically about the x-axis as in Figure 3.14a into two triangles $\triangle PQR$ and $\triangle PQR'$, each having angle $a/2$ at P. Thus applying Proposition 3.14.2 to each gives

$$\text{area } \triangle PRR' = \text{area } \triangle PQR + \text{area } \triangle PQR' = 2(\pi/2 - a/2),$$

as required.

(2) Choose any interior point S of the ideal triangle $\triangle PQR$ with 3 ideal vertexes, and draw in the 3 hyperbolic line segments PS, QS, RS. These subdivide $\triangle PQR$ into 3 triangles $\triangle SPQ, \triangle SQR, \triangle SRP$ of the type considered in (1), as on Figure 3.14c. If a, b, c are the angles at S in each of these, then

$$\text{area } \triangle PQR = \text{area } \triangle SPQ + \text{area } \triangle SQR + \text{area } \triangle SRP$$

$$= \pi - a + \pi - b + \pi - c,$$

which gives what I want, in view of $a + b + c = 2\pi$.   QED

Proof of Theorem 3.14   Starting from a finite triangle $\triangle PQR$, extend sides RP, QR and PQ to infinity to get Figure 3.14d. Now the whole triangle has area equal to π by (2) of the lemma, and it is subdivided into $\triangle PQR$ plus three triangles with two ideal vertexes which have areas a, b, c by (1) of the lemma. Thus the area of $\triangle PQR$ is $\pi - a - b - c$.   QED

**3.14.4
An
alternative
sketch proof**
The above proof depended on an explicit integration. This dependence can be substantially reduced, by an elegant argument making more systematic use of the additivity of angle sums. The alternative is due to David Epstein (who acknowledges hints from C. F. Gauss and N. I. Lobachevsky).

Lemma 1   *Given any two ideal triangles $\triangle PQR$ and $\triangle P'Q'R'$ having three ideal vertexes, there is a Lorentz transformation $A \colon \mathcal{H}^2 \to \mathcal{H}^2$ taking $\triangle PQR$ into $\triangle P'Q'R'$.*

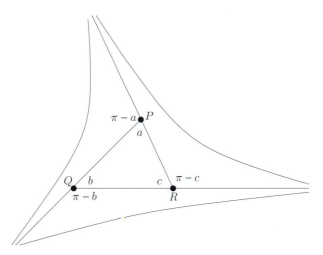

Figure 3.14d   The angular defect formula.

This is an easy exercise in linear algebra: given any three distinct lines V_1, V_2, V_3 of $\mathbb{R}^3$ contained in the cone $\{q_L(\mathbf{v}) = 0\}$, there is a Lorentz basis $\mathbf{e}_0, \mathbf{e}_1, \mathbf{e}_2$ of $\mathbb{R}^3$ for which

$$V_1 = \langle \mathbf{e}_0 + \mathbf{e}_2 \rangle, \quad V_2 = \langle \mathbf{e}_0 + \mathbf{e}_1 \rangle, \quad V_3 = \langle \mathbf{e}_0 - \mathbf{e}_2 \rangle. \tag{10}$$

Lemma 2   *Any ideal triangle $\triangle PQR$ with three ideal vertexes at infinity has finite area π.*

It follows by Lemma 1 that all ideal triangles are congruent, so the key point is that the area is finite (the π can be viewed as an arbitrary scaling factor). There is a beautiful axiomatic geometry proof due to Gauss in Coxeter [5], Figure 16.4a.

Now consider an ideal triangle $\triangle PQR$ with $P \in \mathcal{H}^2$, and two ideal vertexes Q, R. Let $a = \angle QPR$, and write $\triangle PQR = \triangle(a)$. I wish to prove that area $\triangle PQR = \pi - a$. For this purpose, define $L(a) = \pi -$ area $\triangle PQR$.

Lemma 3   *$L(a)$ is an additive function of a, that is, if $a = b + c$ with $0 < a, b, c < \pi$ then $L(a) = L(b) + L(c)$.*

Proof   Immediate from Figure 3.14e:

area $\triangle OPQ +$ area $\triangle OQR =$ area $\triangle OPR +$ area $\triangle PQR =$ area $\triangle OPR + \pi$,

since all vertexes of $\triangle PQR$ are ideal.   QED

Lemma 4   *$L(a)$ is a monotonic function of a, that is, if $a > b$ then $L(a) > L(b)$. Moreover, $L(0) = 0$ and $L(\pi) = \pi$.*

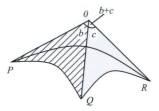

Figure 3.14e   Area is an additive function.

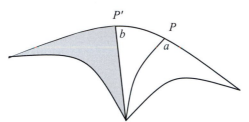

Figure 3.14f   Area is a monotonic function.

Proof   There are several ways of proving that $a > b$ in a figure such as Figure 3.14f consisting of two ideal triangles: if $a \leq b$ then the lines out of P and P' diverge, as discussed in Theorem 3.13. Note that as $a \to 0$ the triangle $\triangle(a)$ tends to the whole of the ideal triangle, and as $a \to \pi$ it tends to a line.   QED

It is obvious that Lemmas 3 and 4 imply that $L(a) = a$, so that area $\triangle(a) = \pi - a$ for all $a \in (0, \pi)$. The proof then concludes as before by referring to Figure 3.14d.

Exercises

In Exercises 3.1–3.10, consider the geometry of the sphere $S^2 \subset \mathbb{R}^3$ of radius 1 with the intrinsic (spherical) metric.

3.1   (a) Define, by analogy with Euclidean geometry, the notions of *spherical circle* and
           spherical disc with centre $P \in S^2$ and radius ρ.
       (b) Prove that a spherical circle with radius $\rho < \pi$ has circumference $2\pi \sin \rho$.
       (c) Prove that a spherical disc of radius $\rho < \pi$ has area $2\pi (1 - \cos \rho)$.
       [Hint: for (c), integrate (b).]

3.2   Deduce from Exercise 3.1 that there does not exist an isometric map from any region
       of S^2 to a region of the Euclidean plane $\mathbb{R}^2$.

3.3   (a) State and prove Pons Asinorum (1.16.1) in spherical geometry.
       (b) Let $P_1, P_2 \in S^2$ be distinct points. Prove that the set of points equidistant from
           P_1, P_2 is a spherical line (great circle). [Hint: use the ambient metric of $\mathbb{R}^3$ to find
           the locus, and (i) to prove in terms of the intrinsic geometry of S^2 that every point
           equidistant from P_1, P_2 is on it.]

3.4   Let $\triangle \subset S^2$ be a spherical n-gon, with internal angles $a_1, \ldots, a_n$ at its vertexes.
       Guess and prove a formula for the area of $\triangle$ in terms of $\sum a_i$. (Assume that the figure

$\triangle$ does not overlap itself to avoid complicated explanations of how you count the area.)

3.5   Let α, β, γ be the side lengths of a spherical triangle $\triangle PQR$ and a, b, c the opposite angles. Use the main formula

$$\cos \alpha = \cos \beta \cos \gamma - \sin \beta \sin \gamma \cos a$$

to prove that $|\beta - \gamma| < \alpha < \beta + \gamma$ and $\alpha + \beta + \gamma < 2\pi$.

  Prove that every triple with α, β, $\gamma < \pi$ satisfying the above inequalities are the sides of a spherical triangle.

3.6   In the same notation, prove the sine rule for spherical triangles

$$\frac{\sin \alpha}{\sin a} = \frac{\sin \beta}{\sin b} = \frac{\sin \gamma}{\sin c}.$$

[Hint: using the notation p, q, r for the vertexes of $\triangle PQR$ as in 3.2, prove that the matrix with rows p, q, r has determinant $\det(p, q, r) = \sin a \sin \beta \sin \gamma$.]

3.7   Prove that if $\triangle$ is an acute angled spherical triangle whose angles are submultiples $\pi/p, \pi/q, \pi/r$ of π, then

$$(p, q, r) = (2, 2, n) \quad \text{or} \quad (2, 3, 3) \quad \text{or} \quad (2, 3, 4) \quad \text{or} \quad (2, 3, 5).$$

Prove that if $\triangle$ is a triangle in $\mathbb{R}^2$ with the same properties, then the possibilities are

$$(p, q, r) = (3, 3, 3) \quad \text{or} \quad (2, 4, 4) \quad \text{or} \quad (2, 3, 6).$$

[Hint: using the formula area $\triangle = a + b + c - \pi$, get $\frac{1}{p} + \frac{1}{q} + \frac{1}{r} > 1$.]

3.8   Show that in polar coordinates

$$x = r \cos \theta, \quad y = r \sin \theta, \quad z = \sqrt{1 - r^2}$$

on the sphere S^2 of unit radius, the element of area in S^2 is

$$dA = \frac{r \, dr \, d\theta}{\sqrt{1 - r^2}}.$$

[Hint: consider a small sector $[\theta, \theta + \delta\theta] \times [r, r + \delta r]$ in $\mathbb{R}^2$. Prove that the sector of S^2 lying over it is very close to a spherical rectangle with length of sides equal to $r \, \delta\theta$ and $\delta r/\sqrt{1 - r^2}$.]

3.9   Here is a general project: take any result you know in plane Euclidean geometry, find an analogue for spherical geometry, and either prove or disprove it. As concrete exercises, prove or deny the following:
(a) the 3 medians of a triangle intersect in a point G;
(b) the 3 perpendicular bisectors of a triangle intersect in a point O;
(c) (harder) the 3 heights of a triangle intersect in a point H.

3.10 Another general project: set up definitions and notation for the geometry of the n-dimensional sphere S^n. [Hint: the ambient space is $\mathbb{R}^{n+1}$ and the distance function comes from the Euclidean inner product.] State and prove some theorems in this more general setting in analogy with the treatment of Chapter 1; in particular, if you feel brave, you can classify completely motions of the 3-sphere S^3 following 1.15.

In Exercises 3.11–3.21, consider the geometry of hyperbolic plane $\mathcal{H}^2$ with the hyperbolic metric.

3.11 Hyperbolic distance is defined by $d(P, Q) = \operatorname{arccosh}(-\mathbf{v} \cdot_L \mathbf{w})$. Adapt the argument of the proof of Theorem B.3 (3) to prove directly that $-\mathbf{v} \cdot_L \mathbf{w} \geq 1$ for $\mathbf{v}, \mathbf{w} \in \mathcal{H}^2$.

3.12 Prove that $P(s) = (\cosh s, \sinh s)$ is the parametrisation of the hyperbola
$$\mathcal{H}^1 : (-t^2 + x^2 = -1) \subset \mathbb{R}^2$$
by arc length in the Lorentz pseudometric $q = -t^2 + x^2$; put more simply, $P(s + ds) - P(s)$ is ds times a vector tangent to Q at $P(s)$ of unit length for q. [Hint: if $P(s) = (\cosh s, \sinh s)$ then $\frac{dP}{ds} = (\cosh s, \sinh s)$, a unit space-like vector.]

3.13 (a) Let $P = (1, 0, 0) \in \mathcal{H}^2$; show how to parametrise the circle centre P and radius $r < \pi$ in $\mathcal{H}^2 \subset \mathbb{R}^3$. Deduce that a circle of radius r has circumference $2\pi \sinh r$; and that a disc with centre P of radius $r < \pi$ has area $2\pi(1 + \cosh r)$. Your formulas should be analogous to those for $S^2 \subset \mathbb{R}^3$ in Exercise 3.1.

(b) Deduce from (a) that there does not exist an isometric map from any region of $\mathcal{H}^2$ to a region of the Euclidean plane $\mathbb{R}^2$ or of the sphere S^2.

(c) A Pringle's potato chip is a reasonably accurate model in Euclidean 3-space of a hyperbolic disc of radius $r = 1$ (isometrically embedded). What happens if we try to make one of radius $r = 100$?

3.14 Define a reflection of $\mathcal{H}^2$, and prove properties analogous to those of reflections of $\mathbb{R}^2$: there exists a reflection taking P_1 to P_2, any direct motion of $\mathcal{H}^2$ is a composite of 2 reflections, any opposite motion is a composite of 3 reflections, Pons Asinorum, etc. [Hint: follow the spherical case in Exercise 3.3.]

3.15 (a) Use the main formula
$$\cosh \alpha = \cosh \beta \cosh \gamma - \sinh \beta \sinh \gamma \cos a$$
to prove that in a right-angled hyperbolic triangle, the hypotenuse is longer than either of the other two sides. If $L \subset \mathcal{H}^2$ is a line and $P \in \mathcal{H}^2$ a point not on L, deduce that the length of the perpendicular dropped from P to L (if it exists) is the shortest distance from P to L.

(b) Consider the function $d(P, Q)$ for $Q \in L$; prove that $d(P, Q)$ takes a minimum value. [Hint: fix attention to a suitable closed ball around P and use the fact that a function on a closed interval attains its bounds.] Deduce that a perpendicular from P to L exists and is unique.

(c) If $L, M \subset \mathcal{H}^2$ are lines not meeting in $\mathcal{H}^2$ and not ultraparallel, prove that L and M have a unique common perpendicular.

3.16 Interpret the matrixes
$$\begin{pmatrix} \cosh s & \sinh s & 0 \\ \sinh s & \cosh s & 0 \\ 0 & 0 & 1 \end{pmatrix} \quad \text{and} \quad \begin{pmatrix} \cosh s & \sinh s & 0 \\ \sinh s & \cosh s & 0 \\ 0 & 0 & -1 \end{pmatrix},$$
as hyperbolic translation and glide.

3.17   In Figure 3.13, let $Q = (1, 0, 0)$ and $P = (t_0, 0, y_0)$, so that the matrix $\begin{pmatrix} t_0 & 0 & y_0 \\ 0 & 1 & 0 \\ y_0 & 0 & t_0 \end{pmatrix}$
defines a hyperbolic translation taking Q to P. Show that the line $L : (y = 0)$ goes to
$M : (t_0 y = y_0 t)$, and the line $y = (\tan \varphi) x$ through Q at angle φ to L (parametrised
by $(t, (\cos \varphi) r, (\sin \varphi) r)$ with $-t^2 + r^2 = -1$) goes to $(\sin \varphi) x = (\cos \varphi)(y_0 t - t_0 y)$.
Conclude that the limiting angle θ in Theorem 3.13 is given by $\cot \theta = y_0$.

3.18   (Harder) The formula $\cosh \alpha = \cosh \beta \cosh \gamma$ gives the hypotenuse α of a right-
angled hyperbolic triangle in terms of the other two sides β, γ. Prove that this is
always longer than the corresponding Euclidean result $\sqrt{\beta^2 + \gamma^2}$.

3.19   Let α, β, γ be the sides (lengths) of a hyperbolic triangle $\triangle PQR$ and a, b, c the
opposite angles. Prove the hyperbolic sine rule

$$\frac{\sinh \alpha}{\sin a} = \frac{\sinh \beta}{\sin b} = \frac{\sinh \gamma}{\sin c}.$$

[Hint: argue as in 3.10 and Exercise 3.6.]

3.20   The hyperbolic lines $L_c : (y = c(t - x))$ with $|c| < 1$ are ultraparallel, tending to
$(1, 1, 0)$ at infinity (see Definition 3.12). Verify that L_c is parametrised by arc length
as

$$L_c : P_c(s) = \left(t_0 e^{-s} + \frac{1}{t_0} \sinh s, \ \frac{1}{t_0} \sinh s, \ y_0 e^{-s} \right),$$

where $y_0 = \frac{c}{\sqrt{1-c^2}}$ and $t_0 = \frac{1}{\sqrt{1-c^2}}$ (so that $c = \frac{y_0}{t_0}$ and $P_c(0) = (t_0, 0, y_0) \in L_c$). Calculate $d(P_c(s), P_{-c}(s))$ and show that the two curves $L_{\pm c}$ approach asymptotically as
$s \to \infty$.

Since $L_0 : (y = 0)$ is sandwiched between $L_{\pm c}$ for any c (e.g. $c = 1/2$), it follows
that L_0 and L_c are asymptotically close. (But you have to start the parametrisation by
arc length at an appropriate point to make the two parametrised curves converge.)

3.21   Suppose that L_1 and L_2 are divergent hyperbolic lines as in Definition 3.12. Set up a
parametrisation by arc length as $L_1 : P(s)$, $L_2 : P'(s)$ and prove that $d(P(s), P'(s))$
must grow at least linearly in the variable s.

3.22   (a) Show that in polar coordinates

$$x = r \cos \theta, \quad y = r \sin \theta, \quad t = \sqrt{1 + r^2},$$

the element of area in $\mathcal{H}^2$ is

$$dA = \frac{r \, dr \, d\theta}{\sqrt{1 + r^2}}.$$

[Hint: consider a small sector $[\theta, \theta + \delta \theta] \times [r, r + \delta r]$ in the space-like Euclidean
$\mathbb{R}^2$. Prove that the sector of $\mathcal{H}^2$ lying over it is very close to a hyperbolic rectangle
with length of sides equal to $r \, \delta \theta$ and $\delta r / \sqrt{1 + r^2}$.]

(b) By writing down the Jacobian determinant for the change of coordinates, check
that the element of area in $\mathcal{H}^2$ in the usual coordinates (t, x, y) is

$$dA = \frac{dx \, dy}{t}.$$

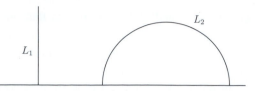

Figure 3.15 $\mathcal{H}$-lines.

The final set of exercises 3.23–3.26 aim to give an alternative *model* of hyperbolic geometry, which may help you visualise some of its properties. I set up a geometry on the complex upper half-plane $\mathcal{H}$ (Exercise 3.23), show that it is the same geometry as the hyperbolic plane $\mathcal{H}^2$ (Exercise 3.24), and investigate the failure of the parallel postulate in the new model (Exercise 3.25). If you want to read further on this, look at Beardon [2], Chapter 7.

3.23   Let

$$\mathcal{H} = \{z = x + iy \in \mathbb{C} \mid y > 0\}$$

be the *upper half-plane* in the complex plane. Define $\mathcal{H}$-*lines* to be of two kinds (see Figure 3.15): either vertical Euclidean half-lines $L_1 = \{x + iy \in \mathcal{H} \mid x = c\}$ for a real constant c, or half-circles $L_2 = \{x + iy \in \mathcal{H} \mid (x - a)^2 + y^2 = c^2\}$ with centre $(a, 0)$ on the real axis $\{y = 0\}$.
Show, algebraically or by drawing pictures, that
(a) two $\mathcal{H}$-lines meet in at most one point;
(b) every pair of distinct points of $\mathcal{H}$ lies on a unique $\mathcal{H}$-line.

3.24   (a) Consider the map φ defined by

$$\varphi : (T, X, Y) \mapsto \left(\frac{-Y + i}{T - X}\right).$$

Show that if $(T, X, Y) \in \mathcal{H}^2$ then $T - X > 0$ hence φ is a map from the hyperbolic plane $\mathcal{H}^2 \subset \mathbb{R}^{2,1}$ to the upper half-plane $\mathcal{H}$.
(b) Consider the map ψ defined by

$$\psi : (x + iy) \mapsto \left(\frac{1 + x^2 + y^2}{2y}, \frac{-1 + x^2 + y^2}{2y}, \frac{-x}{y}\right).$$

Show that if $x + iy \in \mathcal{H}$ then its image $(T, X, Y) \in \mathcal{H}^2$ hence ψ is a map from $\mathcal{H}$ to $\mathcal{H}^2$.
(c) Show that ϕ and ψ are inverse bijections between $\mathcal{H}$ and $\mathcal{H}^2$.
(d) Show that the image of a hyperbolic line $L \in \mathcal{H}^2$ is an $\mathcal{H}$-line and conversely.
(e) Let $z_1, z_2 \in \mathcal{H}$ be points of the upper half-plane, and let $\mathbf{v}_i = \psi(z_i)$ be their images under ψ. Show, using the formulas above, that

$$-\mathbf{v}_1 \cdot_L \mathbf{v}_2 = 1 + \frac{|z_1 - z_2|^2}{2 \operatorname{Im}(z_1) \operatorname{Im}(z_2)}.$$

Deduce that setting

$$d_{\mathcal{H}}(z_1, z_2) = \operatorname{arccosh}\left(1 + \frac{|z_1 - z_2|^2}{2\operatorname{Im}(z_1)\operatorname{Im}(z_2)}\right),$$

makes $(\mathcal{H}, d_{\mathcal{H}})$ into a metric space isometric to $(\mathcal{H}^2, d_{\mathcal{H}^2})$.

Therefore $\mathcal{H}$ has a metric geometry, isometric to the hyperbolic plane $\mathcal{H}^2$. In particular, it has its own symmetries, the $\mathcal{H}$-motions. Sketch some cases like the hyperbolic translations and reflections on a sheet of paper, starting from their geometric definitions. As a matter of fact, any direct $\mathcal{H}$-motion is of the form

$$z \mapsto \frac{az + b}{cz + d}$$

for a real matrix

$$\begin{pmatrix} a & b \\ c & d \end{pmatrix}$$

with $ad - bc > 0$; indirect motions are given by

$$z \mapsto \frac{a(-\bar{z}) + b}{c(-\bar{z}) + d}.$$

If you feel brave, try your hand at proving that these maps preserve $\mathcal{H}$ and its metric; consult Beardon [2], section 7.4 for the full story.

One further point deserves special mention: although there appear to be two different types of $\mathcal{H}$-lines, the set of $\mathcal{H}$-motions acts transitively on the set of $\mathcal{H}$-lines. This holds because the analogous statement is true in $\mathcal{H}^2$, and the two are the same!

3.25 (Graphical exercise) Draw a point $P \in \mathcal{H}$ and an $\mathcal{H}$-line L not containing P. (To make your picture pretty, choose L to be a half-circle and P to be lying over its centre; of course you know that all configurations are like that up to $\mathcal{H}$-motions!) Draw some lines through P meeting L. Shade the region of $\mathcal{H}$ covered by lines through P meeting L. Draw the ultraparallel lines (see Definition 3.12) to L from P. For educational purposes, repeat the exercise with L a 'vertical' line. Now stare at your drawings and contemplate the vast regions in hyperbolic space not contained in lines incident with P and L, as opposed to the case of $\mathbb{E}^2$ where this set is a line.

3.26 (Another graphical exercise) Do Exercise 3.15 (b–c) on $\mathcal{H}$ without any computation, by drawing the appropriate diagrams.

4 Affine geometry

Affine geometry is the geometry of an n-dimensional vector space together with its *inhomogeneous linear structure*. Accordingly, this chapter covers basic material on linear geometries and linear transformations. The inhomogeneous linear maps that we allow as transformations of affine space include translations such as $(x, y) \mapsto (x + a, y + b)$, dilations such as $(x, y) \mapsto (2x, 2y)$ and 'shear' maps such as $(x, y) \mapsto (x, x + y)$. It is impossible to define an origin, distances between points, or angles between lines in a way which makes them invariant under these transformations, or to compare ratios of distances in different directions. However, the line PQ through two points P and Q of $\mathbb{A}^n$ makes perfectly good sense; this is also called the affine span $\langle P, Q \rangle$ of P and Q. An affine line is a particular case of an *affine linear subspace* $E \subset \mathbb{A}^n$; I can view an affine linear subspace as the affine span $\langle P_1, \ldots, P_k \rangle$ of a finite set of points, or as the set of solutions of a system of inhomogeneous linear equations $M\mathbf{x} = \mathbf{b}$. Arbitrary affine linear maps take affine linear subspaces into one another, and also preserve collinearity of points, parallels and ratios of distances along parallel lines; all of these are thus well defined notions of affine geometry.

4.1 Motivation for affine space

As before, I write $\mathbb{R}^n$ for the set of n-tuples $(x_1, \ldots, x_n)$ of real numbers and $V \cong \mathbb{R}^n$ for an n-dimensional vector space over $\mathbb{R}$. The rest of this chapter discusses the same set under the name of affine n-space $\mathbb{A}^n$; Chapter 1 called it Euclidean n-space $\mathbb{E}^n$. Before giving the formal definitions, let me explain briefly the point of having so many alternative names and notations for what are basically all the same thing.

The set $\mathbb{R}^n$ of n-tuples $(x_1, \ldots, x_n)$ is an n-dimensional vector space over the field $\mathbb{R}$ of real numbers: I can add two n-tuples and multiply an n-tuple by a real number. These notions have a physical meaning: in mechanics, for example, you could think of adding vectors in a parallelogram of forces or velocities. A vector space V is the abstract structure in which the operations of linear algebra make sense: addition of vectors and multiplication of vectors by scalars are defined in V, and satisfy some rules. Once I know that V has dimension n, I can choose a basis $\{\mathbf{e}_1, \ldots, \mathbf{e}_n\}$ and identify a vector

$$\mathbf{v} = \sum_{i=1}^{n} x_i \mathbf{e}_i \in V$$

with the n-tuple $(x_1, \ldots, x_n)$, so that $V = \mathbb{R}^n$. However, there may be practical or theoretical reasons for not wanting to fix a basis at the outset: a proof, or the answer to a calculation, may turn out to be much nicer in a well chosen basis. In mechanics, for example, you might want to distinguish forces in the direction of motion from forces perpendicular to the motion.

Similarly, working in coordinate geometry of $\mathbb{R}^n$ (even $\mathbb{R}^2$, of course), there may be reasons to choose coordinates

$$x_i' = x_i - a_i \quad \text{for } i = 1, \ldots, n \tag{1}$$

centred at some point $P = (a_1, \ldots, a_n)$. In mechanics, for example, if two particles at points P and Q exert forces on one other, you may want to take either P or Q as the origin of coordinates, or you may prefer to take their centre of gravity, or some other point. The coordinate change (1) is however not a linear map or change of basis of the vector space V; for example, it has the effect of changing the origin of coordinates to make $P = (0, \ldots, 0)$. Indeed, two different choices of origin differ by a translation of the form (1). Just as the laws of physics should not depend on the choice of origin, we require that geometric properties of affine space are invariant under affine coordinate changes, which include maps of the form (1).

The same issue commonly arises, from a slightly different point of view, in problems where we are interested in some space that is clearly linear in some sense, but has no preferred origin. The model case is the space of solutions to a system of inhomogeneous linear equations $A\mathbf{x} = \mathbf{b}$: as you know, the space of all solutions is given by a *particular* solution $\mathbf{x}_0$ plus the *general solution of the homogenised equations* $A\mathbf{x} = 0$ (the kernel of the matrix A). Solutions of the homogeneous linear equations form a vector space; the particular solution $\mathbf{x}_0$ provides an identification of the set of all solutions with a vector space U. There is no preferred particular solution $\mathbf{x}_0$, and a different particular solution $\mathbf{x}_0'$ gives another identification of the solution set with U, differing by a translation as in (1), with $\mathbf{a} = \mathbf{x}_0' - \mathbf{x}_0$.

4.2 Basic properties of affine space

This section lists basic properties that I take as the definition of affine space $\mathbb{A}^n$.

(I) Affine space has a set of points $P \in \mathbb{A}^n$ in one-to-one correspondence with position vectors $\mathbf{p} \in V$ in an n-dimensional vector space V over $\mathbb{R}$. The one-to-one correspondence $P \leftrightarrow \mathbf{p}$ between points and vectors is not fixed; rather, I am always allowed to translate it by a fixed vector $\mathbf{b}$, so that the new identification is $P \leftrightarrow \mathbf{p}' = \mathbf{p} + \mathbf{b}$.

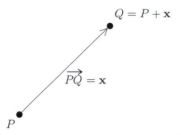

Figure 4.2   Points, vectors and addition.

(II)   Further, a choice of basis of V leads to an identification $V = \mathbb{R}^n$, and thus to a *coordinate system* on $\mathbb{A}^n$, in which points $P \in \mathbb{A}^n$ are represented by coordinates

$$P \leftrightarrow \mathbf{p} = \begin{pmatrix} x_1 \\ \vdots \\ x_n \end{pmatrix} \in \mathbb{R}^n \quad \text{where } x_i \in \mathbb{R}.$$

(III)   Two points $P, Q \in \mathbb{A}^n$ determine a vector $\overrightarrow{PQ} \in V$ as in Figure 4.2. This vector is independent of the identifications discussed in (I).

(IV)   Conversely, a vector $\mathbf{x} \in V$ can be added to a point $P \in \mathbb{A}^n$ to get a new point $Q = P + \mathbf{x} \in \mathbb{A}^n$, and then $\overrightarrow{PQ} = \mathbf{x}$; see again Figure 4.2. This operation is also independent of the identifications discussed in (I).

As with the definition of $\mathbb{E}^n$ in 1.3, the definition of $\mathbb{A}^n$ involves an identification $\mathbb{A}^n = V$ or $\mathbb{A}^n = \mathbb{R}^n$, followed by the assurance that any other identification would do just as well provided that it is related to the first by a suitable transformation, in this case an affine linear transformation. How to define affine space in abstract algebra (without explicit mention of any origin or coordinates) is a slightly arcane issue, and is discussed in 9.2.4.

Remarks   In most of what follows, you can replace $\mathbb{R}$ by other fields. The most obviously useful case is an n-dimensional vector space over $\mathbb{C}$, giving rise to $\mathbb{A}^n_{\mathbb{C}}$, but affine geometries over finite fields $\mathbb{F}_{p^n}$, or over other fields, also have applications in many areas of math and science. I do not intend to labour this point, because doing it properly would involve a lot of algebra of fields, and because the course is directed more towards metric geometries, which are 'real' subjects.

Note also that I work here from the outset in a finite dimensional space V. However, in many areas of math, affine spaces appear as the set of solutions of inhomogeneous linear equations in infinite dimensional spaces: there is no preferred solution, but the differences $\mathbf{x} - \mathbf{x}'$ between any two solutions form a vector space (finite dimensional or otherwise). This happens, for example, in solving $Dx(t) = y(t)$ for functions $x = x(t)$ in a suitable space of differentiable functions, where D is a linear differential operator and $y(t)$ a given function. The spaces of functions we work in, and sometimes also our affine space of solutions, are often infinite dimensional.

4.3 The geometry of affine linear subspaces

An *affine linear subspace* $E \subset \mathbb{A}^n$ is a nonempty subset of the form

$$E = P + U = \{P + \mathbf{v} \mid \mathbf{v} \in U\},$$

with $P \in \mathbb{A}^n$ and $U \subset V$ a vector subspace. By Proposition (1) below, any point of E will do equally well in place of P, so there is no unique origin specified in E.

Let $P, Q \in \mathbb{A}^n$ be two distinct points. The *line* spanned by P and Q is

$$PQ = \{P + \lambda \overrightarrow{PQ} \mid \lambda \in \mathbb{R}\}.$$

The definition clearly shows that PQ is an affine linear subspace, with U the one dimensional vector subspace of V generated by $\overrightarrow{PQ} \in V$. As in 1.2, we have the *line segment* or *interval*

$$[P, Q] = \{P + \lambda \overrightarrow{PQ} \mid 0 \leq \lambda \leq 1\}.$$

It is useful to spell this out in vector notation. If $P, Q \in \mathbb{A}^n$ correspond to position vectors $\mathbf{p}, \mathbf{q}$, their affine span is the set

$$\mathbf{pq} = \{\mathbf{p} + \lambda(\mathbf{q} - \mathbf{p}) \mid \lambda \in \mathbb{R}\} = \{(1 - \lambda)\mathbf{p} + \lambda\mathbf{q} \mid \lambda \in \mathbb{R}\}.$$

The latter is the form of the linear span construction most commonly used. The line segment now becomes

$$[\mathbf{p}, \mathbf{q}] = \{(1 - \lambda)\mathbf{p} + \lambda\mathbf{q} \mid 0 \leq \lambda \leq 1\},$$

as shown in Figure 4.3a.

Three points P, Q, R are *collinear* if they lie on the same line. If I represent the points by position vectors $\mathbf{p}, \mathbf{q}, \mathbf{r}$, this means that $\mathbf{r} = (1 - \lambda)\mathbf{p} + \lambda\mathbf{q}$; as we saw in 1.2, there are three subcases here:

$$\left.\begin{array}{c} \lambda \leq 0 \\ 0 \leq \lambda \leq 1 \\ 1 \leq \lambda \end{array}\right\} \iff \left\{\begin{array}{l} \mathbf{p} \in [\mathbf{r}, \mathbf{q}] \text{ so } P \in [R, Q] \\ \mathbf{r} \in [\mathbf{p}, \mathbf{q}] \text{ so } R \in [P, Q] \\ \mathbf{q} \in [\mathbf{p}, \mathbf{r}] \text{ so } Q \in [P, R]. \end{array}\right.$$

Proposition

(1) *Let $E = P_0 + U$ be an affine linear subspace of $\mathbb{A}^n$. Then the vector space U is uniquely defined by E; explicitly*

$$U = \{\overrightarrow{PQ} \mid P, Q \in E\}.$$

In other words, $E = P + U$ for any $P \in E$.

(2) *A necessary and sufficient condition for a nonempty subset $E \subset \mathbb{A}^n$ to be an affine subspace is that the line PQ is contained in E for all $P, Q \in E$.*

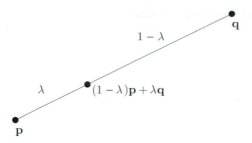

Figure 4.3a   The affine construction of the line segment [**p**, **q**].

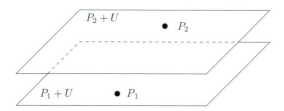

Figure 4.3b   Parallel hyperplanes.

(3)   *A necessary and sufficient condition for E to be an affine subspace is that it is nonempty, and defined by a set of inhomogeneous linear equations in a coordinate system.*

The proofs are easy exercises in linear algebra. (1) states that E can be translated back to the vector space U choosing any point $P \in E$; informally, any point $P \in E$ can serve as origin. (3) spells out the other easy way of specifying an affine linear subspace using coordinates; examples can be found in Exercises 4.1 and 4.5.

Write $\dim E = \dim U$ for the *dimension* of a nonempty affine linear subspace E. The only n-dimensional affine linear subspace is $\mathbb{A}^n$ itself; $\dim E = 0$ means simply that E consists of a single point, whereas a one dimensional affine linear subspace is simply a line. The last interesting case with a name of its own is an affine linear subspace of dimension $n - 1$ (that is, codimension one), a *hyperplane*. Two hyperplanes E_1, E_2 are *parallel*, if they are translates of the same vector subspace of V, that is $E_1 = P_1 + U$, $E_1 = P_2 + U$ with $\dim U = n - 1$, as in Figure 4.3b. An equivalent condition is to ask that the two hyperplanes should either coincide or have no common point.

Definition   Let $\Sigma \subset \mathbb{A}^n$ be any set; an *affine linear combination* of Σ is any point $P \in \mathbb{A}^n$ of the form

$$P = P_0 + \sum_{i=1}^{k} \lambda_i \overrightarrow{P_1 P_i}, \quad \text{where } P_i \in \Sigma \text{ and } \lambda_i \in \mathbb{R}. \tag{2}$$

Using position vectors $\mathbf{p}_i$ of points P_i simplifies this expression once more; an affine linear combination of Σ is any point $P \in \mathbb{A}^n$ of the form

$$\mathbf{p} = \sum_{i=0}^{k} \mu_i \mathbf{p}_i, \quad \text{where } \mathbf{p}_i \in \Sigma \text{ and } \mu_i \in \mathbb{R} \text{ with } \sum_{i=0}^{k} \mu_i = 1. \tag{3}$$

This generalises the expression $(1 - \lambda)\mathbf{p} + \lambda\mathbf{q}$ used to parametrise points of the affine line PQ. The points P_i appear in the form (3) with $(\lambda_0, \ldots, \lambda_k) = (0, \ldots, 1, \ldots, 0)$; this confirms that I really mean $= 1$ in (3) rather than $= 0$.

The *affine span* $\langle \Sigma \rangle$ of any subset is the set of affine linear combinations of Σ. By the previous remark, $\langle \Sigma \rangle$ contains all lines spanned by pairs of points in Σ. If $P \in \Sigma$ then $\langle \Sigma \rangle = P + U$, where $U \subset V$ is the vector subspace spanned by the vectors $\overrightarrow{PQ}$ for $Q \in \Sigma$. Thus $\langle \Sigma \rangle \subset \mathbb{A}^n$ is an affine linear subspace, in fact the smallest one containing all the points of Σ.

4.4 Dimension of intersection

The formula

$$\dim U + \dim W = \dim U \cap W + \dim(U + W) \tag{4}$$

for vector subspaces U, W of a finite dimensional vector space is familiar from linear algebra. You remember the proof: pick a basis of $U \cap W$, extend to two bases of U and W, and the union is a basis of $U + W$.

Theorem   *Let $E, F \subset \mathbb{A}^n$ be affine subspaces. Then*

$$\dim E \cap F = \dim E + \dim F - \dim \langle E, F \rangle, \tag{5}$$

provided that $E \cap F \neq \emptyset$.

The exceptional case $E \cap F = \emptyset$ happens if and only if E, F are contained in parallel hyperplanes. This can happen essentially whatever the dimension of E and F; more precisely, there exist affine linear subspaces E, F with $\dim E = a$, $\dim F = b$, $E \cap F = \emptyset$ and $\dim \langle E, F \rangle = c$ for any $a, b < n$ and any c with

$$\max\{a, b\} + 1 \leq c \leq \min\{n, a + b + 1\}.$$

Proof   The proof of the first statement is almost trivial: if $P \in E \cap F$ then the four affine subspaces in question are translates of the four vector subspaces

$$E', F', E' \cap F', E' + F' \subset V$$

so that the result follows at once from the linear algebra formula (4).

The counterexamples involve affine subspaces E, F of $\mathbb{A}^n$ contained in parallel hyperplanes. To be specific, I choose coordinates and put

$$E \subset \{x_1 = 0\} \quad \text{and} \quad F \subset \{x_1 = 1\}.$$

Then certainly $E \cap F = \emptyset$. The converse is proved in Exercise 4.3.

Assume that $(0, \dots, 0) \in E = E'$ and $(1, 0, \dots, 0) \in F$; then F is the translation by $(1, 0, \dots, 0)$ of a vector subspace $F' \subset V$ contained in $\{x_1 = 0\}$. The equality (4) holds, but the point is that $E \cap F = \emptyset$ takes no account of $\dim E' \cap F'$. Now E' and F' are any vector subspaces contained in the hyperplane given by $x_1 = 0$, so that $\dim E'$, $\dim F'$, $\dim(E' + F')$ can be anything up to and including $n - 1$.   QED

You will find it instructive to spell out the theorem in a few concrete cases. For example, if $n = 2$ and E, F are distinct lines, then $\dim \langle E, F \rangle = 2$ and so the conclusion is that $E \cap F$ is zero dimensional (that is, a point) unless it is empty, the standard dichotomy of intersecting and parallel lines. For $n = 3$, see Exercise 4.2.

4.5   Affine transformations

Recall the following definition, which I repeat here for completeness.

Definition   A map $T : \mathbb{A}^n \to \mathbb{A}^n$ is an *affine transformation* if it is given in a coordinate system by $T(\mathbf{x}) = A\mathbf{x} + \mathbf{b}$, where $A = (a_{ij})$ is an $n \times n$ matrix with nonzero determinant and $\mathbf{b} = (b_i)$ a vector; in more detail,

$$
\mathbf{x} = (x_i) \mapsto \mathbf{y} = \left(\sum_{j=1}^{n} a_{ij} x_j + b_i \right), \quad \text{or} \quad \begin{pmatrix} x_1 \\ \vdots \\ x_n \end{pmatrix} \mapsto A \begin{pmatrix} x_1 \\ \vdots \\ x_n \end{pmatrix} + \begin{pmatrix} b_1 \\ \vdots \\ b_n \end{pmatrix}. \tag{6}
$$

The set $\mathrm{Aff}(n)$ of affine transformations is the set of 'allowed symmetries' of affine space $\mathbb{A}^n$. This set consists of invertible maps from $\mathbb{A}^n$ to $\mathbb{A}^n$ (because I require $\det A \neq 0$). It acts transitively on $\mathbb{A}^n$; that is, a suitable affine transformation maps any point to any other. In particular, there is no distinguished origin, as I said before: every point is like every other. Contrast this with the situation in linear algebra, where the allowed maps $V \to V$ are the *homogeneous* linear maps, all mapping the origin $0 \in V$ to itself.

It is immediate that an affine transformation takes an affine linear subspace to an affine linear subspace; that is, it preserves the incidence geometry of affine linear subspaces. In Proposition 1.9, I proved a converse statement, under the additional assumption that T restricts to an affine linear map on each line. In fact, one can prove that, for $n \geq 2$, a bijective map $T : \mathbb{A}^n \to \mathbb{A}^n$ that preserves lines and is continuous is actually affine linear. (This is a point where working over $\mathbb{R}$ is essential; for a proof, see Exercise 5.22.)

4.6   Affine frames and affine transformations

Definition   A set of points $\{P_0, \dots, P_k\}$ of $\mathbb{A}^n$ is *affine linearly independent* if the k vectors $\overrightarrow{P_0 P_1}, \dots, \overrightarrow{P_0 P_k}$ are linearly independent in V. In other words, a set $\Sigma \subset \mathbb{A}^n$ is *affine linearly dependent* if there exists a nontrivial relation $\sum_{i=0}^{k} \lambda_i \mathbf{p}_i = 0$ between position vectors $\mathbf{p}_0, \dots, \mathbf{p}_k$ of points in Σ, with $\lambda_i \in \mathbb{R}$ and $\sum_{i=0}^{k} \lambda_i = 0$; Σ is *affine linearly independent* if no such relation exists.

A set $\Sigma \subset \mathbb{A}^n$ is an *affine frame of reference* if it is affine linearly independent and spans $\mathbb{A}^n$ (compare the notion of Euclidean frame in 1.12). This means that every point $P \in \mathbb{A}^n$ can be written in the form (2) of 4.3 in a unique way; that is, no proper subset of Σ can span $\mathbb{A}^n$. Equivalently, $\Sigma = \{P_0, P_1, \ldots, P_n\}$ where $P_0 \in \Sigma$ is any point, and the vectors $\overrightarrow{P_0 P_1}, \ldots, \overrightarrow{P_0 P_n}$ form a basis of V.

In view of the correspondence between bases in a vector space and linear maps, the last clause gives the following.

Proposition   *Fix one affine frame of reference $P_0, \ldots, P_n$. Then*

$$T \mapsto T(P_0), \ldots, T(P_n)$$

defines a one-to-one correspondence between affine transformations and affine frames of reference of $\mathbb{A}^n$.

4.7   The centroid

The following proposition is usually thought of as part of (plane) Euclidean geometry; however, it only involves ratios along lines and incidence of lines, so in fact it belongs to affine geometry. The other 'famous' centres of a triangle described in 1.16.4 use notions such as angle or distance that have no meaning in affine geometry.

Proposition   *Let P, Q, R be three points of $\mathbb{A}^n$. Then the three medians of $\triangle PQR$, that is, the three lines connecting each vertex to the midpoint of the opposite side, meet in a common point S.*

Proof   Write $\mathbf{p}, \mathbf{q}, \mathbf{r}$ for the position vectors of P, Q, R. Write $\mathbf{p}' = \frac{1}{2}(\mathbf{q} + \mathbf{r})$ for the midpoint of $\mathbf{q}$ and $\mathbf{r}$ and $\mathbf{s} = \frac{2}{3}\mathbf{p}' + \frac{1}{3}\mathbf{p}$ for the point dividing the segment between $\mathbf{p}$ and $\mathbf{p}'$ in ratio one to two. Then $\mathbf{s} = \frac{1}{3}(\mathbf{p} + \mathbf{q} + \mathbf{r})$ is symmetric in $\mathbf{p}, \mathbf{q}, \mathbf{r}$, so lies on the lines joining $\mathbf{q}$ and $\mathbf{q}' = \frac{1}{2}(\mathbf{p} + \mathbf{q})$ and $\mathbf{r}$ and $\mathbf{r}' = \frac{1}{2}(\mathbf{p} + \mathbf{q})$. Hence the point S with position vector $\mathbf{s}$ lies on all medians of $\triangle PQR$.   QED

To reiterate the point: the statement that this is a theorem of *affine* geometry means that applying any affine transformation takes Figure 4.7 to a figure with the same properties, and in particular takes the centroid of a triangle to the centroid.

Exercises

4.1   Consider the 3 planes

$$\Pi_1 : \{x - 2 = \tfrac{1}{2}(y - z)\}, \quad \Pi_2 : \{x + 2 = y\}, \quad \Pi_3 : \{3(2x + z) = 3y + 1\}$$

in affine space $\mathbb{A}^3$. Calculate $\Pi_1 \cap \Pi_2$ and $\langle \Pi_1, \Pi_2 \rangle$ and find out whether the dimension of intersection formula works; if not, why not? (Compare Theorem 4.4.) Ditto for $\Pi_1 \cap \Pi_3$ and $\langle \Pi_1, \Pi_3 \rangle$.

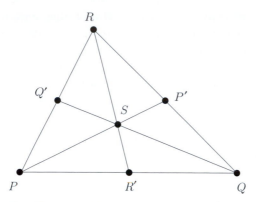

Figure 4.7   The affine centroid.

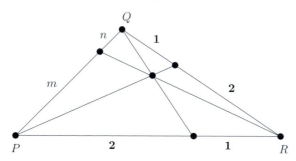

Figure 4.8   A weighted centroid.

4.2   Experiment with 4.4, formula (5) for $n = 3$ and different E, F. For example, classify pairs of lines of $\mathbb{A}^3$ into three types, namely intersecting, parallel and skew, drawing pictures for each case.

4.3   Suppose that E, $F \subset \mathbb{A}^n$ are disjoint affine linear subspaces; prove that there is a linear form φ on $\mathbb{A}^n$ such that $\varphi(E) = 0$ and $\varphi(F) = 1$. [Hint: let $P \in E$, $Q \in F$ and $\mathbf{v} = \overrightarrow{PQ}$. Then $E = P + U$ for a vector subspace $U \subset V$, and $\mathbf{v} \notin U$. Deduce that there exists a linear form on V that is zero on U but nonzero on $\mathbf{v}$.]

4.4   Write down the affine transformation taking

$$(0, 0), (1, 0), (0, 1) \mapsto (2, 1), (5, -1), (3, 8).$$

Can you map the same points $(0, 0)$, $(1, 0)$, $(0, 1)$ to $(2, 1)$, $(5, -2)$, $(3, 0)$ by an affine transformation? Why?

4.5   Determine the dimension of the affine linear subspace E of $\mathbb{A}^5$ given by the equations

$$x_1 + x_3 - 2x_5 = 1$$
$$x_2 - 2x_4 + x_5 = -2$$
$$x_1 + 2x_2 + x_3 - 4x_4 = -3.$$

Find an affine transformation taking E to an affine linear subspace given by $x_1 = \cdots = x_k = 0$ for some value of k. [Hint: choose a suitable affine frame consisting of points on and off the subspaces, compare 4.6.]

4.6   Give a determinantal criterion in coordinates for $n + 1$ points of $\mathbb{A}^n$ to be affine linearly dependent (Definition 4.6). [Hint: start by saying how you tell whether 3 points of $\mathbb{A}^2$ are collinear.]

4.7   In $\triangle PQR$ of Figure 4.8, take points dividing the three sides in the ratios $1 : 2$, $1 : 2$, $n : m$. Assume that the three lines connecting the vertexes to the points on the opposite sides have a common point. Calculate the value of the ratio $m : n$. [Hint: follow the proof of Proposition 4.7. Answer: the ratio is $4 : 1$.]

4.8   A general project: set up affine geometry over the finite field $\mathbb{F}_p$ of integers modulo the prime p. Count the number of points of affine space $\mathbb{A}^n$, and prove analogues of the theorems of the text. Check that everything remains true, with a single exception (harder): the statement concerning the centroid *fails* for one value of p.

5 Projective geometry

The affine geometry studied in Chapter 4 provided one possible solution to the problem of inhomogeneous linear geometry. However, this turns out not to be the only one. This chapter treats the alternative: it introduces projective space $\mathbb{P}^n$ as another equally natural linear geometry. The construction of $\mathbb{P}^n$ can be motivated starting from affine geometry in terms of adding 'points at infinity'.

Projective geometry is simple to study as pure homogeneous linear algebra, ignoring the motivation; 'linear algebra continued' or 'more things to do with matrixes' would be accurate subtitles for this chapter. In $\mathbb{P}^n$, the statement of affine geometry analogous to the dimension of intersection formula of Theorem 4.4 holds without the 'inhomogeneous' conditions of Chapter 4, so that, for example, two distinct lines $L_1, L_2 \subset \mathbb{P}^2$ meet in a point $P = L_1 \cap L_2$ without exception.

Projective geometry has lots of applications in math and other subjects. Projective transformations include the perspectivities, or projections from a fixed viewpoint from one plane to another, that form the foundation of perspective drawing; the fact that you can readily recognise an object from any angle, or a photograph taken from any point (and viewed at any angle) indicates that your brain processes perspectivities automatically and instantaneously.

5.1 Motivation for projective geometry

5.1.1 Inhomogeneous to homogeneous

Recall from Chapter 4 that if E, F are affine linear subspaces of affine space $\mathbb{A}^n$, then there is a nice formula 4.4 expressing the dimension of their intersection *provided* that $E \cap F \neq \emptyset$. One of the points of projective geometry is to get rid of this unpleasant condition. The trouble all comes from the inhomogeneity of the equations: simultaneous inhomogeneous equations include, say, $x_1 = 0$ and $x_1 = 1$, where only two equations reduce $\mathbb{A}^n$ to the empty set.

The solution is the following formal trick. Suppose $\sum a_{ij}x_j = b_i$ is a set of inhomogeneous equations in n unknowns $x_1, \ldots, x_n$ defining an affine linear subspace $E \subset \mathbb{A}^n$. Replace these by homogeneous equations $\sum a_{ij}x_j = b_i x_0$ in $n+1$ unknowns $x_0, x_1, \ldots, x_n$. The solutions with $x_0 \neq 0$ give ratios $x_1/x_0, \ldots, x_n/x_0$ that give a faithful picture of $E \subset \mathbb{A}^n$. But there are also the solutions with $x_0 = 0$, called 'points at infinity'. Including these points adds information to the set of ordinary solutions; namely, information about all the ways the ratios $x_1 : \cdots : x_n$ can behave as the x_i tend to infinity. A solution $0, \xi_1, \ldots, \xi_n$ (with some $\xi_i \neq 0$) corresponds not to a point of E, but to an $(n-1)$-dimensional family of all parallel lines with slope $\xi_1 : \cdots : \xi_n$ satisfying the homogenised equations $\sum a_{ij}\xi_j = 0$, that is, parallel to some line in E (compare Figure 5.8).

The set E together with these extra solutions is a projective linear subspace of projective space; the intersection of projective linear subspaces is then governed by the formula of 4.4 *without exception*. This does not mean that two projective linear subspaces cannot have empty intersection; it only means that they have empty intersection exactly when they have a numerical reason to do so. In modern language, the quantity $\dim E + \dim F - \dim \langle E, F \rangle$ on the right-hand side of formula 4.4 is called the *expected dimension* of the intersection of E and F; in projective geometry, linear subspaces always intersect in a subspace whose dimension equals the expected dimension.

5.1.2
Perspective

You recognise Figure 5.1a as a plane picture of a cube in $\mathbb{R}^3$. The way it is drawn, the horizontal parallel edges appear to meet in points of the plane.

Suppose I fix the origin $O \in \mathbb{A}^3$ and map points of a plane $\Pi \subset \mathbb{A}^3$, to another plane $\Pi' \subset \mathbb{A}^3$ by taking $P \in \Pi$ into the point of intersection $P' = OP \cap \Pi'$ of the line OP with Π'. A map of this kind is called a *perspectivity*. It corresponds to putting your eye at O, with Π' a glass plate, Π behind it with a figure on it, and drawing faithfully the figure on the glass as you see it (see Figure 5.1b).

I get a map $f : \Pi \to \Pi'$ between two planes. It is easy to see that f maps lines of Π to lines of Π', and parallel or concurrent lines L, L', L'', on Π to parallel or concurrent lines M, M', M'' on Π'. Here I am ignoring practicalities, such as the finite extent of the plane represented by a physical piece of glass, or the possibility that some of Π might poke out in front of Π' rather than behind (see Exercise 5.1 for details). Strictly speaking, f is only locally defined, and the conclusions should be qualified by adding 'within the domain of definition'; the activity takes place in the real world, and set theoretic niceties do not cause us undue discomfort.

The map $f : P \mapsto P'$ is constructed in linear terms, but is not actually linear (see Exercise 5.1): choosing coordinates on $\Pi, \Pi' = \mathbb{A}^2$, it can be shown that f is *fractional linear*, that is, of the form

$$f(\mathbf{x}) = \frac{A\mathbf{x} + b}{L\mathbf{x} + c}$$

where A, b, L and c are $2 \times 2, 2 \times 1, 1 \times 2$ and 1×1 matrixes. Note that these can be assembled into a 3×3 matrix $\left(\begin{smallmatrix} A & b \\ L & c \end{smallmatrix} \right)$.

5.1.3
Asymptotes

Figure 5.1c depicts the hyperbola $xy = 1$ and the parabola $y = x^2$. Viewed from a long way off, the hyperbola is very close to the line pair $xy = 0$. In fact, outside a

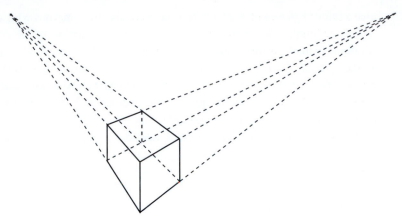

Figure 5.1a   A cube in perspective.

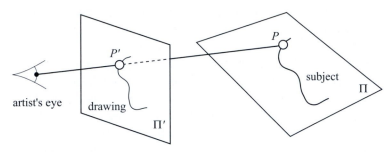

Figure 5.1b   Perspective drawing.

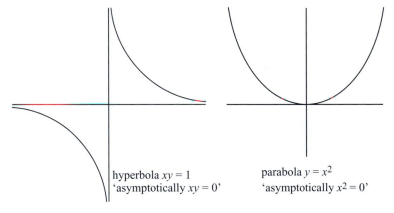

hyperbola $xy = 1$
'asymptotically $xy = 0$'

parabola $y = x^2$
'asymptotically $x^2 = 0$'

Figure 5.1c   Hyperbola and parabola.

big circle of radius R, either $|x| > R$ and $|y| < 1/R$ or vice versa. One can argue that, in turn, the parabola is asymptotic to the line $x = 0$, in the sense that the tangent line at the point (x_0, x_0^2) gets steeper and steeper. This argument is not actually very convincing: when both $x, y \gg 0$, all you can say is $y = x^2 \gg x$. Nevertheless, in the theory of conic sections, it is said, for example, that 'the two branches of the parabola meet at infinity', or that the parabola 'passes through the point at infinity corresponding to lines parallel to $x = 0$.'

The statements on asymptotes are qualitative views of what happens to the curves when x or y is large (quite vague, even arguable for those in quotes). But we have not so far said what asymptotic directions or points at infinity actually are, which is a disadvantage in discussing asymptotes formally or in calculating with them. Making sense of asymptotes (of algebraic plane curves), and providing a simple framework for calculating with them is one thing that projective geometry does very well.

5.1.4 Compact-ification

Here I assume that you know some topology; read this section after Chapter 7 if you prefer.

Affine space $\mathbb{A}^n$ is not compact; in contrast, projective space $\mathbb{P}^n$ is compact, as are its closed subsets, including all projective algebraic varieties. Compact sets are much more convenient than noncompact ones in many contexts of geometry, topology, analysis and algebraic geometry. Given a closed set $X \subset \mathbb{A}^n$, you can compactify it by extending $\mathbb{A}^n$ to $\mathbb{P}^n$; then $X \subset \mathbb{A}^n \subset \mathbb{P}^n$ and the closure $\overline{X} \subset \mathbb{P}^n$ is compact. The points at infinity of the closure $\overline{X}$ correspond in a very precise sense to the asymptotic lines of X, and are calculated by the same simple trick of adding a homogenising coordinate x_0. For example, the hyperbola $xy = 1$ is compactified to the circle S^1 by adding the two points $(\infty, 0)$ and $(0, \infty)$, and the parabola is compactified to S^1 by adding the single point $(0, \infty)$ at which the two branches are said to meet.

5.2 Definition of projective space

Provided you forget about the motivation, the definition is very simple: introduce the equivalence relation $\sim$ on $\mathbb{R}^{n+1} \setminus 0$ defined by

$$(x_0, \ldots, x_n) \sim (y_0, \ldots, y_n) \iff \begin{cases} (x_0, \ldots, x_n) = \lambda(y_0, \ldots, y_n) \\ \text{for some } 0 \neq \lambda \in \mathbb{R}. \end{cases}$$

In other words $\mathbf{x} \sim \mathbf{y}$ if the two vectors $\mathbf{x}$ and $\mathbf{y}$ are proportional, or span the same line (1-dimensional vector subspace) through 0 in $\mathbb{R}^{n+1}$. Then define projective space to be

$$\mathbb{P}^n_{\mathbb{R}} = \mathbb{P}^n = (\mathbb{R}^{n+1} \setminus 0)/\sim = \{\text{lines through 0 in } \mathbb{R}^{n+1}\}.$$

I write $(x_0 : \cdots : x_n)$ for the equivalence class of $(x_0, \ldots, x_n)$; this is the usual notion of relative ratios of $n + 1$ real numbers. $x_0, \ldots, x_n$ are *homogeneous coordinates* on $\mathbb{P}^n$. For example, $\mathbb{P}^1$ is the set of ratios $(x_0 : x_1)$. If $x_0 \neq 0$ you might as well just consider x_1/x_0, but then you are missing one point corresponding to the ratio $(0 : 1)$, where $x_1/x_0 = \infty$.

In coordinate free language, if V is an $(n + 1)$-dimensional vector space over $\mathbb{R}$, write $\mathbb{P}(V)$ for the set of lines of V through 0 (that is, nonzero vectors up to the equivalence $\mathbf{v} \sim \lambda\mathbf{v}$ for $\lambda \neq 0$). Of course, $V \cong \mathbb{R}^{n+1}$ (by a choice of basis), so $\mathbb{P}(V) \cong \mathbb{P}^n$.

A point $P \in \mathbb{P}(V)$ is an equivalence class of vectors $\mathbf{v} \in V$, or a line $\mathbb{R}\mathbf{v}$ through 0; several kinds of notation are popularly used to indicate that $\mathbf{v} = (x_0, \ldots, x_n)$ is a vector in the equivalence class defining P, for example:

$$P = P_\mathbf{v}, \quad P = [\mathbf{v}], \quad \mathbf{v} = \widetilde{P}, \quad P_\mathbf{v} = (x_0 : \cdots : x_n), \quad \text{etc.}$$

To return to the motivation, $\mathbb{P}^n$ contains the subset $(x_0 \neq 0)$ consisting of ratios that can be written $(1 : x_1 : \cdots : x_n)$, which is thus naturally identified with $\mathbb{A}^n$. The language used for motivating projective geometry is quite unsuitable for developing the theory systematically. For example, the terminology of 'points at infinity' is cumbersome and gives a distorted view of the symmetry of the situation.

The formal language of projective geometry is simply a reinterpretation of the ideas of linear algebra; the subset with $x_0 \neq 0$ is not distinguished in $\mathbb{P}^n$, and there is no discrimination against points of the complement (with $x_0 = 0$). Working with the definitions of projective geometry and formal calculations in homogeneous co-ordinates is in many ways easier to understand than how it relates to the motivation discussed in 5.1.1, and I proceed with this, returning to the motivation in 5.8. So for the time being, I discuss the geometry of $\mathbb{P}^n$ in terms of the vector space $\mathbb{R}^{n+1}$, and I advise you to forget the motivation.

5.3   Projective linear subspaces

The only structures enjoyed by $\mathbb{P}(V)$ are derived from V. Thus all statements or calculations for $\mathbb{P}(V)$ must reduce to linear algebra in V and the equivalence relation $\sim$ on points of V.

As a first example, here is the definition of the line PQ through two points $P = (x_0 : \cdots : x_n)$ and $Q = (y_0 : \cdots : y_n)$ of $\mathbb{P}^n$. First lift to $\mathbb{R}^{n+1}$ by setting $\widetilde{P} = (x_0, \ldots, x_n)$ and $\widetilde{Q} = (y_0, \ldots, y_n)$ (that is, pick values of x_i and y_i in the given ratio), then set

$$PQ = \langle P, Q \rangle = \big\{ \text{ratios } (\lambda x_0 + \mu y_0 : \cdots : \lambda x_n + \mu y_n) \text{ for all } (\lambda, \mu) \neq (0, 0) \big\}.$$

The point to notice is that $\lambda P + \mu Q$ is meaningless as a point of $\mathbb{P}^n$, because the ratio $(\lambda x_0 + \mu y_0 : \cdots : \lambda x_n + \mu y_n)$ depends on the choice of $\widetilde{P}$ and $\widetilde{Q}$ within the equivalence classes of P and Q. However, the *set of all* $\lambda \widetilde{P} + \mu \widetilde{Q}$ is a well defined 2-dimensional vector subspace of $V = \mathbb{R}^{n+1}$, and ratios in it form the line PQ.

Thinking in a purely formal way about vector subspaces of a vector space V gives the obvious notion of *projective linear subspace*: if $U \subset V$ is a vector subspace, $\mathbb{P}(U)$ is the subset $(U \setminus 0)/\sim \; \subset \mathbb{P}(V)$ of lines through 0 in U. In other words, if $U \subset \mathbb{R}^{n+1}$ then $\mathbb{P}(U)$ is the set of ratios $(x_0 : \cdots : x_n)$ with $(x_0, \ldots, x_n) \in U$. The dimension of $\mathbb{P}(U)$ is defined to be $\dim \mathbb{P}(U) = \dim U - 1$. Thus $\dim \mathbb{P}^n = n$.

A 0-dimensional subspace is a single point; a 1- or 2-dimensional projective linear subspace is called a *line* or *plane*; an $(n - 1)$-dimensional subspace is a *hyperplane*. I sometimes say *k-plane* to mean k-dimensional projective linear subspace.

Note that the empty set $\emptyset$ is a projective linear subspace: the trivial vector subspace $0 \subset \mathbb{R}^{n+1}$ has $\mathbb{P}(0) = \emptyset \subset \mathbb{P}^n$. By convention we write $\dim \emptyset = -1$, to agree with the general definition just given. As a rule, prudence might suggest that in mathematical arguments, we avoid attaching excessive weight to mumbo-jumbo concerning the empty set or the elements thereof, but here the convention $\dim \emptyset = -1$ has a precise and useful meaning *(in the context of the geometry of linear subspaces only!)*.

Definition   If $\Sigma \subset \mathbb{P}(V)$ is a set, write $\widetilde{\Sigma} \subset V$ for the union of the lines in Σ; let U be the vector subspace of V spanned by $\widetilde{\Sigma}$, and define the *span* or *linear span* of Σ to be $\langle \Sigma \rangle = \mathbb{P}(U)$. This is the smallest projective linear subspace containing Σ.

If $P_0, \ldots, P_s$ are $(s+1)$ points then $\dim \langle P_0, \ldots, P_s \rangle \leq s$; equality holds if and only if the vectors $\widetilde{P}_0, \ldots, \widetilde{P}_s \in \mathbb{R}^{n+1}$ are linearly independent. In this case, $P_0, \ldots, P_s$ are said to be *linearly independent* in $\mathbb{P}^n$.

5.4 Dimension of intersection

Theorem   *Let $E, F \subset \mathbb{P}^n$ be projective linear subspaces. Then*

$$\dim E \cap F = \dim E + \dim F - \dim \langle E, F \rangle; \qquad (1)$$

here the convention $\dim \emptyset = -1$ is in use.

Proof   Write $\widetilde{E}, \widetilde{F} \subset \mathbb{R}^{n+1}$ for the vector subspaces overlying E and F. Then $E \cap F = \mathbb{P}(\widetilde{E} \cap \widetilde{F})$ and $\langle E, F \rangle = \mathbb{P}(\widetilde{E} + \widetilde{F})$. By the linear algebra formula 4.4 (4) we have

$$\dim(\widetilde{E} \cap \widetilde{F}) = \dim \widetilde{E} + \dim \widetilde{F} - \dim(\widetilde{E} + \widetilde{F}), \qquad (2)$$

and since $\dim \mathbb{P}(U) = \dim U - 1$ for every vector subspace $U \subset \mathbb{R}^{n+1}$, (1) follows by subtracting 1 from each term on the left- and right-hand sides of (2).   QED

5.5 Projective linear transformations and projective frames of reference

A nonsingular linear map $\mathbb{R}^{n+1} \to \mathbb{R}^{n+1}$ represented by an invertible matrix A acts in an obvious way on the set of lines of $\mathbb{R}^{n+1}$ through 0: namely, it takes the line $\mathbb{R}\mathbf{v}$ to $\mathbb{R}(A\mathbf{v})$ for every $0 \neq \mathbf{v} \in \mathbb{R}^{n+1}$. A map $T : \mathbb{P}^n \to \mathbb{P}^n$ is a *projective transformation* (also called *projectivity* or *projective linear map*) if it arises in this way from a linear map. In other words, if we write $P_{\mathbf{v}} \in \mathbb{P}^n$ for the point represented by $\mathbf{v} \in \mathbb{R}^{n+1}$, then T is a projective transformation if there is an invertible matrix A such that

$$T(P_{\mathbf{v}}) = P_{A\mathbf{v}} \qquad \text{for all } \mathbf{v} \in \mathbb{R}^{n+1}.$$

Here $A\mathbf{v}$ is the product of A and $\mathbf{v}$, viewed as a column vector. The set of all projective transformations is written $\mathrm{PGL}(n+1)$.

Because $\mathbf{v}$ and $\lambda\mathbf{v}$ represent the same point of $\mathbb{P}^n$, a scalar matrix $\lambda \cdot \mathrm{id} = \mathrm{diag}(\lambda, \ldots, \lambda)$ with $\lambda \neq 0$ acts as the identity. Moreover, if A is an invertible matrix and $\lambda \in \mathbb{R}$ and $\lambda \neq 0$, then A and the product λA have exactly the same effect on every point of $\mathbb{P}^n$. Thus the set of projective transformations is

$$\text{PGL}(n+1) = \{\text{invertible } (n+1) \times (n+1) \text{ matrixes}\}/\mathbb{R}^*$$

where $\mathbb{R}^* = \{\lambda \cdot \text{id} \mid 0 \neq \lambda \in \mathbb{R}\}$.

The following definition, which may seem unexpected at first, is quite characteristic of projective geometry.

Definition   A *projective frame of reference* (or *simplex of reference*) of $\mathbb{P}^n$ is a set $\{P_0, \ldots, P_{n+1}\}$ of $n+2$ points such that any $n+1$ are linearly independent, that is, span $\mathbb{P}^n$.

This means

1.   there exists a basis $\mathbf{e}_0, \ldots, \mathbf{e}_n$ of $\mathbb{R}^{n+1}$ such that $P_i = P_{\mathbf{e}_i}$ for $i = 0, \ldots, n$;
2.   the final point P_{n+1} is $P_{\mathbf{e}_{n+1}}$, where

$$\mathbf{e}_{n+1} = \sum_{i=0}^{n} \lambda_i \mathbf{e}_i, \qquad \text{with } \lambda_i \neq 0 \text{ for every } i.$$

Indeed, the first $n+1$ points $P_0, \ldots, P_n$ are linearly independent, and the final point P_{n+1} is not contained in any of the $n+1$ hyperplanes $\{x_i = 0\}$. The *standard frame of reference* is

$$P_i = (0 : \cdots : 1 : \cdots : 0) \quad (\text{with 1 in the } i\text{th place})$$

$$\text{and } P_{n+1} = (1 : 1 : \cdots : 1).$$

(3)

That is, $\mathbf{e}_i$ for $i = 0, \ldots, n$ is the standard basis of $\mathbb{R}^{n+1}$ and $\mathbf{e}_{n+1} = \sum_{i=0}^{n} \mathbf{e}_i$. The final point $P_{n+1} = (1 : \cdots : 1)$ is there to 'calibrate' the coordinate system.

Theorem   *Let $\{P_0, \ldots, P_{n+1}\}$ be the standard frame of reference. Then there is a one-to-one correspondence between projective transformations and frames of reference, defined by $T \mapsto T(P_0), \ldots, T(P_{n+1})$.*

Proof   Write $\mathbf{e}_0, \ldots, \mathbf{e}_n$ for the standard basis of $\mathbb{R}^{n+1}$, and set $\mathbf{e}_{n+1} = \sum_{i=0}^{n} \mathbf{e}_i$. Now let $\{Q_0, \ldots, Q_{n+1}\}$ be a different frame of reference, and choose representatives $\mathbf{f}_0, \ldots, \mathbf{f}_n, \mathbf{f}_{n+1} \in \mathbb{R}^{n+1}$ of the points $Q_0, \ldots, Q_{n+1}$.

Since $\mathbf{e}_0, \ldots, \mathbf{e}_n$ and $\mathbf{f}_0, \ldots, \mathbf{f}_n$ are two bases of $\mathbb{R}^{n+1}$, the usual result of linear algebra is that there is a uniquely determined linear map $A: \mathbb{R}^{n+1} \to \mathbb{R}^{n+1}$ such that $A\mathbf{e}_i = \mathbf{f}_i$ for $i = 0, \ldots, n$. If $\mathbf{f}_0, \ldots, \mathbf{f}_n$ are column vectors, A is the matrix with the given columns $\mathbf{f}_i$. However, that is not what is given, and not what is required! If you understand that, you have understood the proof.

Indeed, the $\mathbf{f}_i$ are determined only up to scalar multiples. Start again: for any nonzero multiples $\lambda_i \mathbf{f}_i$ of $\mathbf{f}_i$ (for $i = 0, \ldots, n$), there is a uniquely determined linear map $A: \mathbb{R}^{n+1} \to \mathbb{R}^{n+1}$ such that $A\mathbf{e}_i = \lambda_i \mathbf{f}_i$ for $i = 0, \ldots, n$, given by the matrix

A with columns $\lambda_i \mathbf{f}_i$. Using the assumption that $\mathbf{f}_0, \dots, \mathbf{f}_n$ is a basis, I choose the λ_i such that $\mathbf{f}_{n+1} = \sum_{i=0}^{n} \lambda_i \mathbf{f}_i$. Then, because $Q_0, \dots, Q_{n+1}$ is a frame of reference, $\lambda_i \neq 0$ for $i = 0, \dots, n$, and $A\mathbf{e}_{n+1} = \mathbf{f}_{n+1}$ by choice of A. Since $A \colon \mathbb{R}^{n+1} \to \mathbb{R}^{n+1}$ is a linear map with $\mathbf{e}_i \mapsto \lambda_i \mathbf{f}_i$ and $\mathbf{e}_{n+1} \mapsto \mathbf{f}_{n+1}$, it defines a projective linear map $T \colon \mathbb{P}^n \to \mathbb{P}^n$ taking $P_i \mapsto Q_i$ for $i = 0, \dots, n+1$.

For the uniqueness, let us look back through the construction: first, the condition $T(P_i) = Q_i$ for $i = 0, \dots, n$ determines the columns of A up to multiplying each column by a scalar λ_i; so far, any λ_i will do (possibly different choices for different columns). Next, the condition $T(P_{n+1}) = Q_{n+1}$ fixes the λ_i up to a common scalar factor: because we must send $\mathbf{e}_{n+1} = \sum \mathbf{e}_i$ into a multiple of $\mathbf{f}_{n+1} = \sum_{i=0}^{n} \lambda_i \mathbf{f}_i$, we have to choose these values of λ_i. The only remaining choice in A would be to multiply the whole thing through by a scalar. Thus T is uniquely determined.   QED

5.6 Projective linear maps of ℙ¹ and the cross-ratio

Corollary   *There exists a unique projective linear transformation of $\mathbb{P}^1$ taking any 3 distinct points $P, Q, R \in \mathbb{P}^1$ to any other 3.*

Since any 3 distinct points go into any other 3 points, I can say that projective linear transformations act 3-*transitively* on $\mathbb{P}^1$ (Figure 5.6a). This means that there can be no nontrivial function $d(P, Q)$ of 2 points or $\sigma(P, Q, R)$ of 3 points that is invariant under these transformations.

However, there is a function of 4 distinct points invariant under projective linear transformations, namely their *cross-ratio* $\{P, Q; R, S\}$. To define it, note that any choices of representatives $\mathbf{p}, \mathbf{q} \in \mathbb{R}^2 \setminus 0$ of P, Q form a basis. Choosing this basis gives

$$P = (1 : 0), \quad Q = (0 : 1), \quad R = (1 : \lambda) \quad \text{and} \quad S = (1 : \mu) \tag{4}$$

for some λ, μ. Set $\{P, Q; R, S\} = \lambda/\mu$.

Changing the representative $\mathbf{q} \mapsto \mu\mathbf{q}$ sets $\mu = 1$ so that $S = (1 : 1)$. Thus the definition amounts to taking P, Q, S as the frame of reference of $\mathbb{P}^1$, and then defining $\{P, Q; R, S\} = \lambda$, where $R = (1 : \lambda)$. Since by Theorem 5.5, the projective transformation taking P, Q, S to $(1 : 0), (0 : 1), (1 : 1)$ is unique, $\{P, Q; R, S\}$ is well defined, and invariant under transformations in PGL(2).

Remark   To see the point of cross-ratio, it is useful to compare the invariant quantities in $\mathbb{A}^1$ and in $\mathbb{P}^1$. In $\mathbb{A}^1$, to be able to measure, you need to fix the points 0 and 1, then any other point P is fixed by $\lambda = (x - 0)/(1 - 0)$. In $\mathbb{P}^1$ you need also to fix the point at infinity.

Proposition   *Consider four distinct lines of $\mathbb{R}^2$ through $O = (0, 0)$ that are the equivalence classes of P, Q, R, S, and let L be any line of $\mathbb{R}^2$ not through the origin*

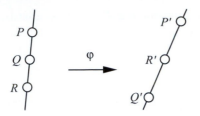

Figure 5.6a   The 3-transitive action of PGL(2) on $\mathbb{P}^1$.

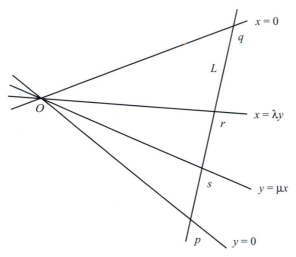

Figure 5.6b   The cross-ratio $\{P, Q; R, S\}$.

intersecting these four lines in $\mathbf{p}, \mathbf{q}, \mathbf{r}, \mathbf{s}$ *respectively (see Figure 5.6b). Then*

$$\{P, Q; R, S\} = \frac{\mathbf{p} - \mathbf{r}}{\mathbf{p} - \mathbf{s}} \cdot \frac{\mathbf{q} - \mathbf{s}}{\mathbf{q} - \mathbf{r}}. \tag{5}$$

Here the quotients on the right-hand side are ratios of vectors along L. You could equally take them as ratios of x-coordinates or y-coordinates of the points; or equally, the ratio of (signed) lengths $\frac{\pm|\mathbf{p}-\mathbf{r}|}{\pm|\mathbf{p}-\mathbf{s}|} \cdot \frac{\pm|\mathbf{q}-\mathbf{s}|}{\pm|\mathbf{q}-\mathbf{r}|}$.

Proof   As in the definition of $\{P, Q; R, S\}$, choose $\mathbf{p}$ and $\mathbf{q}$ as the standard basis of $\mathbb{R}^2$. Then L is given by $x + y = 1$. If λ, μ are as in (4) then $\mathbf{r} \in \mathbb{R}^2$ is in the equivalence class of $(1 : \lambda)$ and is on L, so that necessarily

$$\mathbf{r} = \frac{(1, \lambda)}{1 + \lambda}; \quad \text{similarly} \quad \mathbf{s} = \frac{(1, \mu)}{1 + \mu}. \tag{6}$$

The remaining calculation is very easy:

$$\left.\begin{array}{ll} \mathbf{p} - \mathbf{r} = \frac{\lambda}{1+\lambda}(1, -1), & \mathbf{q} - \mathbf{r} = \frac{-1}{1+\lambda}(1, -1) \\[2mm] \mathbf{p} - \mathbf{s} = \frac{\mu}{1+\mu}(1, -1), & \mathbf{q} - \mathbf{s} = \frac{-1}{1+\mu}(1, -1) \end{array}\right\} \implies \frac{\mathbf{p} - \mathbf{r}}{\mathbf{p} - \mathbf{s}} \cdot \frac{\mathbf{q} - \mathbf{s}}{\mathbf{q} - \mathbf{r}} = \frac{\lambda}{\mu}. \tag{7}$$

This proves the proposition.   QED

5.7 Perspectivities

Let Π, Π' be hyperplanes in $\mathbb{P}^n$ and let O be a point outside Π and Π'. The *perspectivity* $f: \Pi \to \Pi'$ from O is obtained by mapping $P \in \Pi$ to the point of intersection $f(P)$ of the projective line OP with Π'. Note that since O is not on Π', the line OP cannot be contained in Π', and hence the intersection of OP with Π' is a single point by the dimension of intersection formula Theorem 5.4. The case $n = 3$, perspectivity between two planes in 3-space, was described in 5.1.2 and illustrated on Figure 5.1b. As opposed to the example in 5.1.2 (compare Exercise 5.1), the map f is everywhere defined, since new points have been added to affine space to form projective space; this will be discussed further below.

It is easy to write a perspectivity in terms of suitable coordinates. Choose coordinates $(x_0 : x_1 : \cdots : x_n)$ so that $\Pi = \{x_0 = 0\}$, $\Pi' = \{x_1 = 0\}$ and $O = (1, 1, 0, \ldots, 0)$. Then for a point $P = (0 : x_1 : \cdots : x_n)$ of Π, the line OP is the set of points $\{(\lambda : \lambda + \mu x_1 : \mu x_2 : \cdots : \mu x_n)\} \subset \mathbb{P}^n$ (compare the first paragraph of 5.3). The intersection point with Π' is then at $(\lambda : \mu) = (-x_1 : 1)$, so

$$f: (0 : x_1 : \cdots : x_n) \mapsto (-x_1 : 0 : x_2 : \cdots : x_n).$$

In particular, you can view the perspectivity f as a projective transformation from $\Pi = \mathbb{P}^{n-1}$ with coordinates $(x_1 : \cdots : x_n)$ to $\Pi' = \mathbb{P}^{n-1}$ with coordinates $(x_0 : x_2 : \cdots : x_n)$ given by the matrix $A = \mathrm{diag}(-1, 1, \ldots, 1)$.

Proposition *The cross-ratio of four points on a line is invariant under perspectivities; namely, if L is a line in Π and $P, Q, R, S \in L$ are four points on the line, then*

$$\{P, Q; R, S\} = \{f(P), f(Q); f(R), f(S)\}.$$

Proof First of all, the right-hand side of this expression is defined, since the image of L is a line in Π'; this follows from the fact that f is a projective transformation, but as an exercise you can check that it also follows from the definition of f and the dimension of intersection formula. Then $f: L \to f(L)$ is a projective transformation between lines; the cross-ratio is preserved under projective transformations of $\mathbb{P}^1$, so it is preserved under perspectivities also. Note that the equality of cross-ratios also follows from Figure 5.6b and the discussion of Proposition 5.6, once you restrict the discussion to the plane $\mathbb{P}^2 \subset \mathbb{P}^n$ spanned by O and L, and interpret O in Figure 5.6b as a point of this $\mathbb{P}^2$ rather than the affine origin $(0, 0) \in \mathbb{R}^2$.   QED

5.8 Affine space $\mathbb{A}^n$ as a subset of projective space $\mathbb{P}^n$

A hyperplane $H \subset \mathbb{P}^n$ corresponds to an n-dimensional subspace $W \subset \mathbb{R}^{n+1}$, the kernel of a linear form $\alpha: \mathbb{R}^{n+1} \to \mathbb{R}$. Then $\mathbb{P}^n \setminus H$ can be naturally identified with $\mathbb{A}^n$, and $H = \mathbb{P}^{n-1}$ with sets of parallel lines in $\mathbb{A}^n$. The point is very simple: given

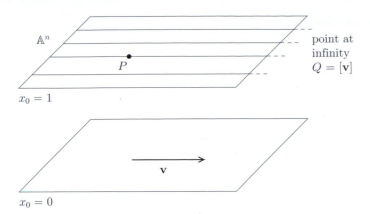

$x_0 = 1$

$x_0 = 0$

Figure 5.8   The inclusion $\mathbb{A}^n \subset \mathbb{P}^n$.

ℓ, I can choose coordinates in $\mathbb{R}^{n+1}$ so that

$$\alpha(x_0, \ldots, x_n) = x_0$$

is the first coordinate. Then

$$\mathbb{P}^n \setminus H = \left\{ \text{ratios } (x_0 : \cdots x_1 : \cdots : x_n) \mid x_0 \neq 0 \right\}$$

$$= \left\{ n\text{-tuples } \left(\frac{x_1}{x_0}, \ldots, \frac{x_n}{x_0} \right) \right\} = \mathbb{A}^n.$$

In Figure 5.8, P is a point with $x_0 \neq 0$, so its equivalence class contains a unique point in the affine hyperplane $\mathbb{A}^n$ defined by ($x_0 = 1$). A point Q with $x_0 = 0$ does not correspond to any actual point of $\mathbb{A}^n$; instead, it corresponds to all the lines of $\mathbb{A}^n$ parallel to $\mathbf{v} = \widetilde{Q}$.

Note that this discussion reverses the process of 'going from inhomogeneous to homogeneous' sketched in 5.1.1; the points of the hyperplane $H \subset \mathbb{P}^n$ are at infinity when viewed from the affine space $\mathbb{A}^n$ defined by ($x_0 = 1$). However, splitting points into 'finite' and 'infinite' is not intrinsic to projective space, but depends on the choice of H (or the linear form α).

5.9   Desargues' theorem

Theorem (Desargues' theorem)   *Let $\triangle PQR$ and $\triangle P'Q'R'$ be 2 triangles in $\mathbb{P}^n$ with $n \geq 2$. Suppose that $\triangle PQR$ and $\triangle P'Q'R'$ are in perspective from some point $O \in \mathbb{P}^n$ (that is, OPP', OQQ' and ORR' are lines). Then the corresponding sides of $\triangle PQR$ and $\triangle P'Q'R'$ meet in 3 collinear points. In other words,*

$$\left. \begin{array}{l} QR \text{ and } Q'R' \text{ meet in } A \\ PR \text{ and } P'R' \text{ meet in } B \\ PQ \text{ and } P'Q' \text{ meet in } C \end{array} \right\} \quad \text{and } A, B, C \text{ are collinear} \tag{8}$$

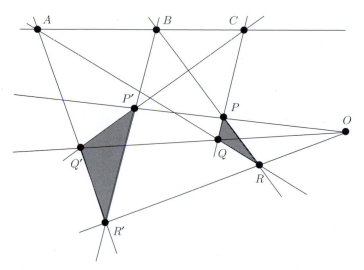

Figure 5.9a   The Desargues configuration in $\mathbb{P}^2$ or $\mathbb{P}^3$.

(see Figure 5.9a). The converse also holds: condition (8) *implies that* $\triangle PQR$ *and* $\triangle P'Q'R'$ *are in perspective from some point* O.

Proof   If the two triangles are in perspective from O, the linear subspaces $\langle O, P, Q, P', Q'\rangle$ and $\langle O, P, R, P', R'\rangle$ are planes that have at least the line $\langle O, P, P'\rangle$ in common. Hence

$$\dim\langle O, P, Q, R, P', Q', R'\rangle = 2 \text{ or } 3$$

by Theorem 5.4. Also, the construction of A, B, C in (8) makes sense: the two lines PQ and $P'Q'$ are coplanar (contained in the plane $\langle O, P, Q, P', Q'\rangle$), so meet in a unique point C, and similarly for the other pairs of sides.

Step 1   Suppose first that P, Q, R and P', Q', R' span a 3-dimensional space $\mathbb{P}^3$ so are not in any $\mathbb{P}^2$, and that they are in perspective from O. Set $L = \langle P, Q, R\rangle \cap \langle P', Q', R'\rangle$. This is the intersection of two distinct planes in $\mathbb{P}^3$, and is therefore a line by Theorem 5.4. But by construction, $A \in L$ since $A = QR \cap Q'R'$. The same applies to B and C, so that also $B, C \in L$ and the 3 points are collinear.

Step 2   We reduce to the first case. Thus suppose that P, Q, R and P', Q', R' are in the plane $\Pi = \langle OPQRP'Q'R'\rangle$. Let $M \in \mathbb{P}^3 \setminus \Pi$ be any point, and lift R, R' off the plane: pick S, S' as in Figure 5.9b in perspective from O such that S and R are in perspective from M, and S' and R' are likewise in perspective from M. Then $\triangle PQS$ and $\triangle P'Q'S'$ are as in Step 1. So the 3 points

$$QS \cap Q'S' = \widetilde{A}, \quad PS \cap P'S' = \widetilde{B} \quad \text{and} \quad PQ \cap P'Q' = C$$

are collinear in $\mathbb{P}^3$, so lie on a line $\widetilde{L} \subset \mathbb{P}^3$. But it is easy to see from the construction that $\widetilde{A}, \widetilde{B}$ lie above A, B in perspective from M, so A, B, C are collinear.

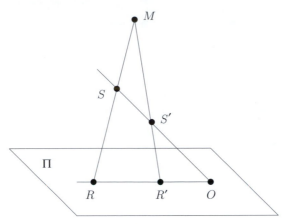

Figure 5.9b   Lifting the Desargues configuration to $\mathbb{P}^3$.

For proofs of the converse see Exercises 5.14–5.15 and 5.11.   QED

It is interesting to note exactly what is used in the proof of Desargues' theorem just given. It is pure *incidence geometry* in $\mathbb{P}^n$ with $n \geq 3$, in the sense that it uses nothing beyond particular cases of formula (1) of Theorem 5.4: two distinct points of $\mathbb{P}^n$ span a line, two concurrent lines span a plane, two distinct lines in a plane intersect in a point, two distinct planes of $\mathbb{P}^3$ intersect in a line, etc. The final part of the proof, Step 2, assumes also that there exists a point not in the plane Π (that is, that we are in $\mathbb{P}^n$ with $n \geq 3$), and that the two lines MR and MR' each have at least one point in addition to M, R and M, R'.

5.10   Pappus' theorem

Theorem (Pappus' theorem)   *Let $L, L' \subset \mathbb{P}^2$ be two lines and*

$$P, Q, R \subset L \quad and \quad P', Q', R' \subset L'$$

two triples of distinct points on L and L' (not equal to $L \cap L'$). Then the 3 points

$$QR' \cap Q'R = A, \quad PR' \cap P'R = B \quad and \quad PQ' \cap P'Q = C$$

are collinear (see Figure 5.10). Notice that the figure is a configuration of 9 lines and 9 points with 3 lines through each point and 3 points on each line.

Proof   This can also be proved via a lifting to $\mathbb{P}^3$, but this requires a bit more information about $\mathbb{P}^3$ (specifically, quadric surfaces in $\mathbb{P}^3$ and properties of lines on them). I sketch the easy proof in coordinates.

By Theorem 5.5, I can choose homogeneous coordinates $(x : y : z)$ such that

$$P = (1 : 0 : 0), \quad Q = (0 : 1 : 0), \quad P' = (0 : 0 : 1) \quad and \; Q' = (1 : 1 : 1).$$

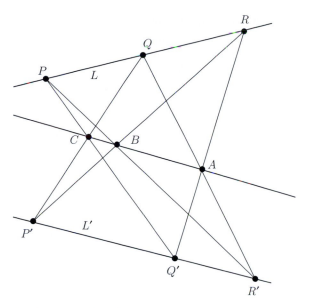

Figure 5.10 The Pappus configuration.

Then

$$L = PQ : \{z = 0\}, \ P'Q : \{x = 0\}, \ L' = P'Q' : \{x = y\} \text{ and } PQ' : \{y = z\}.$$

Therefore

$$C = P'Q \cap PQ' = (0 : 1 : 1).$$

Now let $R = (1 : \beta : 0)$ and $R' = (1 : 1 : \gamma)$. Then easy calculations give

$$PR' : \{z = \gamma y\} \qquad P'R : \{y = \beta x\}$$

so that $B = (1 : \beta : (\beta\gamma))$ and

$$QR' : \{z = \gamma x\} \qquad Q'R : \{y - z = \beta(x - z)\}$$

so that $A = (1 : (\beta + \gamma - \beta\gamma) : \gamma)$. Finally, A, B, C are all on the line

$$\{y - z = \beta(1 - \gamma)x\}. \quad \text{QED}$$

5.11 Principle of duality

Projective duality is based on the idea that the space $(\mathbb{R}^{n+1})^*$ of linear forms $\alpha : \mathbb{R}^{n+1} \to \mathbb{R}$ is also isomorphic to $\mathbb{R}^{n+1}$. Namely, if $\mathbf{e}_0, \dots, \mathbf{e}_{n+1}$ is a basis of $\mathbb{R}^{n+1}$ then the dual basis is given by the linear form

$$\mathbf{e}_i^* : \mathbb{R}^{n+1} \to \mathbb{R} \quad \text{defined by} \quad \mathbf{e}_i^*(\mathbf{e}_j) = \delta_{ij} = \begin{cases} 1 & \text{if } i = j \\ 0 & \text{if } i \neq j. \end{cases}$$

Further, there is a natural one-to-one correspondence between subspaces of $\mathbb{R}^{n+1}$ and its dual: a subspace $V \subset \mathbb{R}^{n+1}$ corresponds to its annihilator (perpendicular) subspace $V^\perp$, that is, the set of linear forms $\alpha: \mathbb{R}^{n+1} \to \mathbb{R}$ vanishing on V. By elementary linear algebra, $\dim V + \dim V^\perp = n + 1$. Hence we obtain the following correspondence between elements of the geometry of projective linear subspaces of $\mathbb{P}^n = \mathbb{P}(\mathbb{R}^{n+1})$ and those of $(\mathbb{P}^n)^* = \mathbb{P}(\mathbb{R}^{n+1})^*$:

$$E = \mathbb{P}(V) = \mathbb{P}^d \subset \mathbb{P}^n \qquad \longleftrightarrow \qquad E^\perp = \mathbb{P}^{n-d-1} = \mathbb{P}(V^\perp) \subset (\mathbb{P}^n)^*$$

$$\text{point } P = \mathbb{P}^0 \in \mathbb{P}^n \qquad \longleftrightarrow \qquad \text{hyperplane } \mathbb{P}^{n-1} = \Pi \subset (\mathbb{P}^n)^*$$

$$\text{subspace } E_1 \subset E_2 \qquad \longleftrightarrow \qquad \text{supspace } E_1^\perp \supset E_2^\perp$$

$$\text{intersection } E_1 \cap E_2 \qquad \longleftrightarrow \qquad \text{span } \langle E_1^\perp, E_2^\perp \rangle$$

$$\text{span } \langle E_1, E_2 \rangle \qquad \longleftrightarrow \qquad \text{intersection } E_1^\perp \cap E_2^\perp.$$

The case of $\mathbb{P}^2$ is special and particularly illustrative: hyperplanes in $\mathbb{P}^2$ are simply lines $L = \mathbb{P}^1 \subset \mathbb{P}^2$; points are dual to lines, and the line through two points is dual to the intersection of two lines.

Proposition (Principle of duality for $\mathbb{P}^2$)   *Every theorem concerning points and lines in $\mathbb{P}^2$ has a dual theorem, obtained from the original one via the following substitutions:*

$$\text{points } P \qquad \longleftrightarrow \qquad \text{lines } L$$

$$\text{lines } L \qquad \longleftrightarrow \qquad \text{points } P$$

$$\text{line } P_1 P_2 \text{ (= the span } \langle P_1, P_2 \rangle) \qquad \longleftrightarrow \qquad \text{point of intersection } L_1 \cap L_2$$

$$\text{intersection } L_1 \cap L_2 \qquad \longleftrightarrow \qquad \text{line } P_1 P_2.$$

This means that given a theorem and its proof about points and lines in $\mathbb{P}^2$, you get a new theorem and its proof by replacing points by lines etc., in a completely automatic way. For example, the dual of Desargues' theorem in $\mathbb{P}^2$ is its converse (which is why I omitted the proof in 5.9). For the dual of Pappus' theorem, see Exercise 5.16.

5.12   Axiomatic projective geometry

An *axiomatic projective plane* Π (Figure 5.12a) consists of two sets

$$\text{Points}(\Pi) \text{ and Lines}(\Pi)$$

and a relation

$$\text{Incidence}(\Pi) \subset \text{Points}(\Pi) \times \text{Lines}(\Pi),$$

usually called an *incidence relation*. If $(P, L) \in \text{Incidence}(\Pi)$, we say that 'point P is on line L' or 'line L passes through point P'; because this is an axiomatic system, we might as well say with David Hilbert 'beer mug P is on table L'.

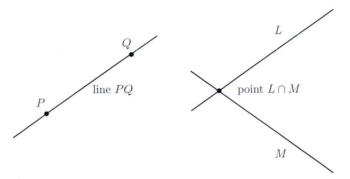

Figure 5.12a Axiomatic projective plane.

This data is subject to the following axioms.

1. Every line has at least 3 points.
2. Every point has at least 3 lines through it.
3. Through any 2 distinct points there is a unique line.
4. Any 2 distinct lines meet in a unique point.

Note that these axioms are obviously dual: you can replace the beer mugs on the tables throughout, and vice versa, and the axioms continue to hold.

More generally an *axiomatic projective space* has a lattice of projective linear subspaces, the incidence relation $\subset$, intersection and linear span, and suitable axioms. It is best not to insist a priori that the dimension of the space or its projective linear subspaces is specified. The most important case is the infinite dimensional case, which von Neumann used to give axiomatic foundations to quantum mechanics, when dimensions of projective linear subspaces can take values in $\mathbb{R}$ or the value ∞.

Introducing coordinates in axiomatic projective planes The real projective plane $\mathbb{P}^2 = \mathbb{P}^2_{\mathbb{R}}$ discussed thus far is certainly not the only axiomatic projective plane: given any field k, you can take $\mathbb{P}^2_k = (k^3 \setminus \{0\})/\sim$ where $(x_0, x_1, x_2) \sim (\lambda x_0, \lambda x_1, \lambda x_2)$ for $0 \neq \lambda \in k$. It is an easy exercise to show that axioms 1 to 4 continue to hold in $\mathbb{P}^2_k$. For example, if $k = \mathbb{F}_2$ you get an axiomatic projective plane with 7 points and 7 lines (see Exercise 5.21).

For this purpose, k has to be a *division ring*, meaning that $ax = b$ has a solution for every $a, b \in A$ with $a \neq 0$, but it is not necessary that k is commutative: you just have to take care that in the equivalence relation $(x_0, x_1, x_2) \sim (\lambda x_0, \lambda x_1, \lambda x_2)$ only *left* multiplication by $\lambda \in k^*$ is allowed, and the linear subspaces of k^3 used to define lines are *right* k-subspaces. Indeed, even the associative law on k can be weakened, although some kind of associativity is required in order that $(x_0, x_1, x_2) \sim (\lambda x_0, \lambda x_1, \lambda x_2)$ is an equivalence relation. For a nontrivial example, do Exercise 8.23. In this course, I do not have time for a detailed discussion of the following result, one of the most beautiful contributions of geometry to pure algebra; for details, consult Hartshorne [12].

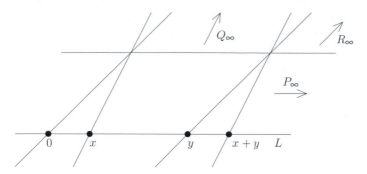

Figure 5.12b   Geometric construction of addition.

Theorem (Hilbert's construction)   *An axiomatic projective plane Π gives rise to a division ring A such that $\Pi = \mathbb{P}_A^2$. Moreover,*

$$A \text{ is an associative ring} \iff \text{Desargues' theorem holds in } \Pi;$$

$$A \text{ is a commutative ring} \iff \text{Pappus' theorem holds in } \Pi.$$

Flavour of proof   We must make a number of choices in Π. Pick a line L_∞ to serve as the line at infinity, three points P_∞, Q_∞ and R_∞ on it, and a line L through P_∞, distinct from L_∞. The elements of the division algebra A are the points of L except for $\infty = L \cap L_\infty$. Now pick 2 different points of $L \setminus \infty$, and call them 0 and 1. The algebraic operation $+$ is constructed in terms of parallels (since we have fixed L_∞, two lines of Π are *parallel* if their intersection is on L_∞) and $\times$ in terms of similarity. For example, addition is defined as in Figure 5.12b.

Exercises

5.1   Let x, y, z be coordinates in $\mathbb{R}^3$, and $\Pi : (z = 1)$, $\Pi' : (y = 1)$ two hyperplanes. Write down the perspectivity $\varphi : \Pi \to \Pi'$ from $O = (0, 0, 0)$ in terms of coordinates (x, y) on Π and (x, z) on Π'. Find and describe the points of Π where φ is not defined. Prove that φ takes a line $L \subset \Pi$ to a line $L' = \varphi(L) \subset \Pi'$ (with a single exception). Consider the pencil of parallel lines $y = mx + c$ of Π (for m fixed and c variable), and determine how φ maps.

5.2   In the notation of the preceding exercise, let $S : (x^2 + y^2 = 1) \subset \Pi$. Understand the effect of the perspectivity φ on S, both geometrically and in coordinates. Show that a circle and a hyperbola in $\mathbb{R}^2$ correspond to projectively equivalent curves in $\mathbb{P}_\mathbb{R}^2$. Account for the 4 asymptotic directions of the hyperbola in terms of S.

5.3   In $\mathbb{P}^2$, write down the equation of the line joining $P = (1 : 1 : 0)$ and $(\alpha : 0 : \beta)$; write down the point of intersection of the 2 lines $x + y + z = 0$ and $\alpha x + \beta y = 0$.

5.4   Let $\Pi_i \subset \mathbb{R}^3$ be the 3 planes of Exercise 4.1. Construct $\mathbb{P}^3$ by introducing a fourth coordinate t, write down the planes of $\mathbb{P}^3$ by homogenising the equations of Π_i, and calculate again the intersections and spans.

5.5 Prove that 3 lines L, M, N of $\mathbb{P}^n$ that intersect in pairs are either concurrent (have a common point) or coplanar. [Hint: use dimension of intersection.]

5.6 Suppose that L, M, N are 3 lines of $\mathbb{P}^4$ not all contained in any hyperplane. Prove that there exists a unique line meeting all 3 lines. [Hint: consider first the span $\langle L, M \rangle = \mathbb{P}^3$.]

5.7 Write down all the projective linear maps φ of $\mathbb{P}^2$ taking

$$(1 : 0 : 0) \mapsto (1 : 2 : 3), \quad (0 : 1 : 0) \mapsto (2 : 1 : 3), \quad (0 : 0 : 1) \mapsto (3 : 1 : 2).$$

Now write down the unique projective linear map taking the standard frame of reference

$$(1 : 0 : 0), \quad (0 : 1 : 0), \quad (0 : 0 : 1), \quad (1 : 1 : 1)$$

into

$$(1 : 2 : 3), \quad (2 : 1 : 3), \quad (3 : 1 : 2), \quad (1 : 2 : 2)$$

respectively. [Hint: reread the proof of Theorem 5.5.]

5.8 Consider the affine linear map $\varphi_0 \colon \mathbb{A}^2 \to \mathbb{A}^2$ given by

$$(x, y) \mapsto (3x - 2, 4y - 3).$$

Prove that φ_0 has a unique fixed point in $\mathbb{A}^2$. [Hint: you can do this by linear algebra, or by using the contraction mapping theorem from metric spaces.]

Write down the projective linear map φ of $\mathbb{P}^2$ extending φ_0. Find the locus of fixed points of φ on $\mathbb{P}^2$. [Hint: either find the fixed points 'by observation', or prove that $(x : y : z)$ is a fixed point of a projective linear map $\mathbf{x} \mapsto A\mathbf{x}$ if and only if $\mathbf{x} = (x, y, z)$ is an eigenvector of A.]

5.9 Repeat the previous question for the map

$$(x, y) \mapsto (x - y + 2, x + y + 3).$$

5.10 Suppose $z = (1 - \lambda)x + \lambda y$. Write $y = (1 - \lambda')x + \lambda'z$; find λ' as a function of λ. Similarly, determine the effect of each permutation of x, y, z on the affine ratio $\lambda = (z - x)/(y - x)$. Thus permuting the 3 points x, y, z defines an action of the symmetric group S_3 on the set of values of λ.

5.11 Let $P, Q, R = (1 : 0), (0 : 1), (1 : 1)$ be the standard frame of reference of $\mathbb{P}^1$.
   (a) Find the projective linear map that takes P, Q, S to Q, P, S (in that order); next P, Q, S to P, S, Q. What is the effect of your map on the affine coordinate of a point $R = (1 : \lambda) \in \mathbb{P}^1$?
   (b) Verify that the matrixes $\left(\begin{smallmatrix} 0 & 1 \\ 1 & 0 \end{smallmatrix} \right)$ and $\left(\begin{smallmatrix} 1 & -1 \\ 0 & -1 \end{smallmatrix} \right)$ generate a group under matrix multiplication isomorphic to the symmetric group S_3.

(c) The cross-ratio of 4 points $\mathbf{p}, \mathbf{q}, \mathbf{r}, \mathbf{s}$ on a line is defined to be

$$\{\mathbf{p}, \mathbf{q}; \mathbf{r}, \mathbf{s}\} = \frac{\mathbf{p} - \mathbf{r}}{\mathbf{p} - \mathbf{s}} \cdot \frac{\mathbf{q} - \mathbf{s}}{\mathbf{q} - \mathbf{r}}.$$

Explain what happens when $\mathbf{p}, \mathbf{q}, \mathbf{r}, \mathbf{s}$ are permuted. Prove that there are in general 6 values

$$\lambda, \ \frac{1}{\lambda}, \ 1 - \frac{1}{\lambda}, \ \lambda - \frac{1}{\lambda}, \ \frac{1}{1 - \lambda}, \ \frac{\lambda}{1 - \lambda}$$

for the cross-ratio, and the group fixing one value is a 4-group V_4.

5.12   Deduce Proposition 1.16.3 (1) from the invariance of cross-ratio under perspectivity. [Hint: interpret one of the four lines in Proposition 5.6 as the line at infinity.]

5.13   Desargues' theorem 5.9 states that if $\triangle PQR$ and $\triangle P'Q'R'$ are 2 triangles in perspective from a point then the 3 points of intersection (e.g., $C = PQ \cap P'Q'$) of corresponding sides are collinear. See Figure 5.9a. Give the coordinate proof. [Hint: as in the proof of Theorem 5.10, take 4 of the points as frame of reference, choose convenient notation for the 3 remaining points, find the coordinates of A, B, C and prove they are collinear.]

5.14   Modify the argument to prove the converse of Desargues' theorem.

5.15   State and prove the dual of Desargues' theorem. Use the same Figure 5.9a.

5.16   State and prove the dual of Pappus' theorem. [Hint: with care you can choose notation exactly dual to that in 5.10, e.g., $p : (x = 0)$, $L = p \cap q = (0 : 0 : 1)$, etc.]

5.17   State and prove the dual of the statement of Exercise 5.6. [Hint: ... given three 2-planes of $\mathbb{P}^4$ not ...]

5.18   Do the same for Exercise 5.5.

5.19   Let $L, L' \subset \mathbb{P}^2$ be two lines. Prove that a projective linear map $\varphi : L \to L'$ can be written as the composite of at most 2 perspectivities $L \to M$ and $M \to L'$ from suitably chosen points of $\mathbb{P}^2$. [Hint: Step 1. If the point of intersection $L \cap L' = P$ is mapped to itself by φ, show that φ is a perspectivity because you can fix the centre O to deal with 3 points. Step 2. In general, choose a third line M and a centre O so that φ composed with the perspectivity $\psi : M \to L$ is as in Step 1.]

5.20   Prove that $\mathbb{P}^n$ has a decomposition as a disjoint union of $n + 1$ subsets

$$\mathbb{P}^n = \{pt\} \sqcup \mathbb{A}^1 \sqcup \mathbb{A}^2 \sqcup \cdots \sqcup \mathbb{A}^n.$$

[Hint: $\mathbb{P}^n = \mathbb{A}^n \sqcup$ hyperplane at ∞.]

5.21   If k is a finite field with q elements, find 2 different proofs of

$$\#(\mathbb{P}^n_k) = 1 + q + q^2 + \cdots + q^n.$$

[Hint: the 'topological' proof uses the decomposition of the preceding exercise. The 'arithmetic' method just counts using the definition $\mathbb{P}^n_k = (k^{n+1} \setminus 0)/k^*$.]

5.22   Prove the following statement, announced in 4.5. For $n \geq 2$, a bijective map $T : \mathbb{A}^n \to$
       $\mathbb{A}^n$, which preserves the incidence geometry of affine linear subspaces of $\mathbb{A}^n$ and is
       continuous, is affine linear. [Hint: it is clearly sufficient to restrict to $n = 2$. Use the
       idea of the sketch proof of Hilbert's theorem 5.12 to show that any such map is affine
       linear, possibly composed with a continuous field automorphism of $\mathbb{R}$. Conclude by
       showing that $\mathbb{R}$ has no nontrivial continuous field automorphisms.]

6 Geometry and group theory

The substance of this chapter can be expressed as the slogan

Group theory is geometry and geometry is group theory.

In other words, every group is a transformation group: the only purpose of being a group is to *act on a space*. Conversely, geometry can be discussed in terms of transformation groups. Given a space X and a group G made up of transformations of X, the geometric notions are quantities measured on X which are invariant under the action of G. This chapter formalises the relation between geometry and groups, and discusses some geometric issues for which group theory is a particularly appropriate language.

   The action of a transformation group on a space is another way of saying *symmetry*. To say that an object has symmetry means that it is taken into itself by a group action: rotational symmetry means symmetry under the group of rotations about an axis. As a frivolous example, Coventry market pictured in Figure 6.0 has (approximate) rotational symmetry: if you stand at the centre, all directions outwards are virtually indistinguishable; you can understand a coordinate frame as a signpost to break the symmetry, and to enable people to find their way around.

   Each of the geometries studied in previous chapters had transformations associated with it: Euclidean motions of $\mathbb{E}^2$, orthogonal transformations as motions of S^2, Lorentz transformations as motions of $\mathcal{H}^2$, and affine and projective linear transformations of $\mathbb{A}^n$ and $\mathbb{P}^n$. In each case, the transformations form a group. I have already studied aspects of this setup: for example, several theorems state that transformations are uniquely determined by their effect on a suitable coordinate frame.

   Whenever two branches of mathematics relate in this way, both can benefit from the cooperation. The repercussions of symmetry extend into many areas of math and other sciences. Some examples:

1.   The basic idea of the Galois theory of fields is to view the roots of a polynomial as permuted amongst themselves by the symmetry group of a field extension.
2.   Crystallography makes essential use of group theory to understand and classify the symmetries of lattice structures formed by crystals, and their impurities.

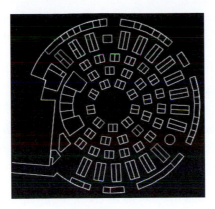

Figure 6.0   The plan of Coventry market.

3. Requiring the laws of physics to be invariant under a symmetry group has been one of the most fertile sources of new ideas in math physics:

    (a) The assumption in Newtonian dynamics that the laws of motion are invariant under Euclidean changes of inertial frames leads directly to conservation of momentum and angular momentum; this will be discussed further in 9.3.1.

    (b) The fact that Maxwell's equations of electromagnetism are not invariant under the Galilean group of symmetries of classical Newtonian dynamics, but are invariant under Lorentzian symmetries, led Einstein to the idea of special relativity.

    (c) Modern particle physics classifies elementary particles in terms of irreducible representations of symmetry groups. Several particles were first predicted from a knowledge of group representations, before being discovered experimentally. (See 9.3.3–9.3.4 for more details.)

    (d) In general relativity, Einstein's field equation for the curvature tensor of spacetime was discovered as the only possible partial differential equation invariant under the pseudo-group of local diffeomorphisms. Einstein himself understood a great deal more about the principles underlying symmetry in physics than about curvature in Riemannian geometry.

We divide math up into separate areas (analysis, mechanics, algebra, geometry, electromagnetism, number theory, quantum mechanics, etc.) to clarify the study of each part; but the equally valuable activity of integrating the components into a working whole is all too often neglected. Without it, the stated aim of 'taking something apart to see how it ticks' degenerates imperceptibly into 'taking it apart to ensure it never ticks again'.

6.1 Transformations form a group

A *transformation* of a set X is a bijective map $T : X \to X$. (We could equally well say *permutation* of X, although this is mainly used for finite sets.) If T is bijective, then so is its inverse T^{-1}. If T_1 and T_2 are maps from X to itself then, as discussed in 2.1, the *composite* $T_2 \circ T_1$ means 'T_2 follows T_1' or 'first do T_1, then do T_2'. If T_1 and T_2 are bijective then so is $T_2 \circ T_1$; thus composition $\circ$ is a binary operation

$$\text{Trans } X \times \text{Trans } X \to \text{Trans } X,$$

where Trans X is just the set of all transformations of X.

Proposition   *Transformations of a set X form a group* Trans X, *with composition of maps as the group operation,* $\text{id}_X : X \to X$ *as the neutral element and* $T \mapsto T^{-1}$ *as the inverse.*

Proof   This is absolutely content free, but let us check the group axioms anyway.

Associative   As discussed in 2.4, $T_3 \circ T_2 \circ T_1$ has no meaning other than the map $X \to X$ taking $x \mapsto T_3(T_2(T_1(x)))$, so that composition of maps is associative.

Inverse $T \circ T^{-1} = T^{-1} \circ T = \text{id}_X$. By definition $T^{-1}(x) = y$ if and only if $T(y) = x$. So $T(T^{-1}(x)) = T(y) = x$ and $T^{-1}(T(y)) = T^{-1}(x) = y$.

Identity $\text{id}_X \circ T = T \circ \text{id}_X = T$. The left-hand side says 'first do T, then do nothing'. In view of which, you might as well omit the second step.   QED

6.2   Transformation groups

A *transformation group* is a subgroup of Trans X for some set X. In other words, it is a subset $G \subset \text{Trans } X$ of bijections $T : X \to X$, containing id_X, and closed under composition $T_1, T_2 \mapsto T_2 \circ T_1$ and inverse $T \mapsto T^{-1}$.

Discussion   Usually X has extra structures (for example: distance, algebraic structure, collinearity structure, topology, distinguished elements or subsets), and we take the set of transformations that preserve these structures:

$$G = \left\{ T \in \text{Trans } X \mid T \text{ preserves the given structures of } X \right\}.$$

It will usually be obvious that

$$T \text{ preserves structures} \implies \text{so does } T^{-1};$$
$$T_1, T_2 \text{ preserve structures} \implies \text{so does } T_2 \circ T_1; \tag{1}$$

so that we get for free that G is a subgroup. This notion includes the symmetry group of an object, automorphisms in algebra, and many other notions you will meet later in math and other subjects.

Example 1.   'No structure'   Let X be a finite set containing n elements labelled $\{1, \ldots, n\}$. The *symmetric group* S_n is the group of all permutations of X.

Example 2. Euclidean motions Motions of $\mathbb{E}^n$ form a group Eucl(n). You can verify this by using the result that a motion T is of the form $T(\mathbf{x}) = A\mathbf{x} + \mathbf{b}$, and write out the composition and inverse in this form (compare 2.2). However, this is completely unnecessary: the result is a standard consequence of what I just said, because motions are defined explicitly as transformations that preserve distance, so that (1) holds. The group Eucl(n) has a subgroup consisting of elements T fixing a chosen point $P \in \mathbb{E}^n$; if P is the origin, then $T(\mathbf{x}) = A\mathbf{x}$ with A an orthogonal matrix. Hence this subgroup is isomorphic to the *orthogonal group* O(n) of $n \times n$ real orthogonal matrixes. (See 6.5.3 for more on this point.)

Example 3. Symmetry groups Let S be a subset of Euclidean space $\mathbb{E}^n$, and let G be the set of isometries of $\mathbb{E}^n$ which map points of S to points of S. Again, the general discussion implies that G is a group, since it is the set of transformations of $\mathbb{E}^n$ preserving the metric and points of S. G is called the *symmetry group of S*. To get interesting groups, one chooses special S (see Exercises 6.5–6.6); for a 'potato-shaped' set S, there will be no nontrivial symmetries at all.

Example 4. Linear maps If V is a vector space over the reals, a transformation $T : V \to V$ is linear if and only if $T(\lambda\mathbf{x} + \mu\mathbf{y}) = $ what you think; that is, T preserves the vector space structure. Thus invertible linear transformations form a group GL(V), the *general linear group* of V. If V has finite dimension, a basis in V gives an identification $V = \mathbb{R}^n$; invertible linear maps are then represented by $n \times n$ invertible matrixes which form the *general linear group* GL($n, \mathbb{R}$). Closely related to the group GL($n + 1, \mathbb{R}$) is the *projective linear group* PGL(n) of projective transformations discussed in 5.5.

We will see that many of the results of the previous chapters, and many other questions at the heart of geometry, can be stated as properties of groups such as Eucl(n), GL(V) or PGL(V).

6.3 Klein's Erlangen program

Around 1870, Felix Klein formulated the following meta-definition:

Geometry is the study of properties invariant under a transformation group.

I have used this principle throughout the previous chapters; for example, distances and angles are geometric properties in Euclidean geometry exactly because they are invariant under motions.

In this context, consider the chain

$$\text{Euclidean geometry } \mathbb{E}^n \to \text{affine geometry } \mathbb{A}^n \to \text{projective geometry } \mathbb{P}^n. \quad (2)$$

The corresponding groups of transformations can be expressed as an increasing chain

$$\text{Eucl}(n) \subset \text{Aff}(n) \subset \text{PGL}(n + 1).$$

Here the inclusion of $\mathrm{Aff}(n)$ as a subgroup of $\mathrm{PGL}(n+1)$ results from the inclusion $\mathbb{A}^n \subset \mathbb{P}^n$ as the set of points with $x_0 \neq 0$: writing $T \in \mathrm{Aff}(n)$ as usual in the form $T(x_1, \ldots, x_n) = A\mathbf{x} + \mathbf{b}$ gives

$$
t \begin{pmatrix} x_1 \\ \vdots \\ x_n \\ 1 \end{pmatrix} = \begin{pmatrix} A & \mathbf{b} \\ 0 & 1 \end{pmatrix} \begin{pmatrix} x_1 \\ \vdots \\ x_n \\ 1 \end{pmatrix},
$$

so that $T \in \mathrm{Aff}(n)$ corresponds to $\left(\begin{smallmatrix} A & \mathbf{b} \\ 0 & 1 \end{smallmatrix} \right)$. It is clear that an element of $\mathrm{PGL}(n+1)$ is in $\mathrm{Aff}(n)$ if and only if it takes the hyperplane $\{x_0 = 0\}$ into itself.

The Erlangen program explains the relation between the three geometries in (2) by saying that as the transformation group gets larger, the invariant properties become fewer: Euclidean geometry has distances and angles; these are no longer invariants of affine geometry, but $\mathbb{A}^n$ has parallels and ratios of parallel vectors; neither of these notions survives in $\mathbb{P}^n$. As I said in 5.6, the action of the projective group $\mathrm{PGL}(2)$ on $\mathbb{P}^1$ is 3-*transitive*, and it is precisely the size of this symmetry group that says that there can be no distance function $d(P, Q)$ of two points, and no ratio of distances $d(P, Q) : d(P, R)$ along lines defined in projective geometry. The group action was prominently involved in the definition of the cross-ratio in 5.6 and in the deduction that it is a well defined function of 4 collinear points.

6.4   Conjugacy in transformation groups

In general, let X be a set and $G \subset \mathrm{Trans}\, X$ a transformation group of X as in 6.1. Suppose that $T \in G$ is a transformation we want to study, and $g \in G$ any element.

Question   What is the conjugate element gTg^{-1}?

Answer gTg^{-1} is just T viewed from a different angle. We can think of gTg^{-1} as acting on elements $gx \in gX$, rather than $x \in X$, by the rule $gx \mapsto g(Tx)$. In fact, the calculation is not very difficult:

$$
gTg^{-1}(gx) = gT(gg^{-1})x = g(T(x)). \tag{3}
$$

Thus we can think of g as a 'change of view', and gTg^{-1} as T expressed in the new view. In many cases, g will actually be a change of basis in a vector space, and gTg^{-1} the same map T written out in terms of the new basis.

Example 1.   Transpositions in S_n   Consider the transposition (12) in the symmetric group S_n of all permutations of $\{1, \ldots, n\}$, the element which transposes 1 and 2 and leaves everything else fixed. Let $g \in S_n$ be any permutation. Then by what I just said, $g(12)g^{-1}$ should also be a transposition, because it is just (12) viewed from another

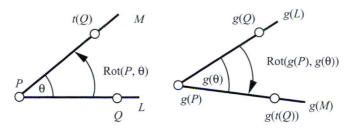

Figure 6.4a   The conjugate rotation $g \operatorname{Rot}(P, \theta) g^{-1} = \operatorname{Rot}(g(P), g(\theta))$.

angle. In fact

$$g(12)g^{-1} = (ab), \quad \text{where } a = g(1), b = g(2).$$

Proof   I give the proof, at the risk of spelling out the really obvious:

$$g(12)g^{-1}: \quad \begin{matrix} g(1) \mapsto 1 \mapsto 2 \mapsto g(2), \\ g(2) \mapsto 2 \mapsto 1 \mapsto g(1), \end{matrix} \qquad (4)$$

and if $c \neq g(1), g(2)$ then $g^{-1}(c) \neq 1, 2$ so that (12) fixes it, and therefore $c \mapsto g^{-1}(c) \mapsto$ itself $\mapsto c$.   QED

Example 2.   Fixed point   Finding the fixed point (or fixed points) of a transformation is an important issue in many geometric contexts. If T fixes P then gTg^{-1} fixes $g(P)$. The calculation is again really obvious, see (3).

Example 3.   Rotation   Let $T = \operatorname{Rot}(P, \theta)$ be a rotation of $\mathbb{E}^2$ and $g \in \operatorname{Eucl}(2)$ any motion. I determine gTg^{-1}. In order to see the action, consider any line L through P, and let M be the line such that $\angle LPM = \theta$. Then T is determined as taking a point $Q \in L$ into the corresponding point of M (that is, $T(Q)$ is the same distance along M).

Now, as I said, we should view gTg^{-1} as acting on $g(\mathbb{E}^2)$. So draw $g(P)$, $g(L)$ and $g(M)$. Then gTg^{-1} fixes $g(P)$, and takes points of $g(L)$ into the corresponding points of $g(M)$ (see Figure 6.4a). This shows that $gTg^{-1} = \operatorname{Rot}(g(P), g(\theta))$, where I write $g(\theta)$ for the angle $\angle g(L)g(P)g(M)$; in fact $g(\theta) = \pm\theta$ (according as g is direct or opposite).

Example 4.   Translation   Let $T : \mathbb{A}^n \to \mathbb{A}^n$ be the translation $\mathbf{x} \mapsto \mathbf{x} + \mathbf{b}$ and suppose that $g \in \operatorname{Aff}(n)$ is given by $\mathbf{x} \mapsto A\mathbf{x} + \mathbf{c}$. By what I said, there is only one thing gTg^{-1} could possibly be – please guess it before reading further.

Now g^{-1} is given by $\mathbf{y} \mapsto A^{-1}(\mathbf{y} - \mathbf{c})$. So gTg^{-1} is the map

$$\mathbf{y} \mapsto A^{-1}(\mathbf{y} - \mathbf{c}) \mapsto A^{-1}(\mathbf{y} - \mathbf{c}) + \mathbf{b} \mapsto A\big(A^{-1}(\mathbf{y} - \mathbf{c}) + \mathbf{b}\big) + \mathbf{c}. \qquad (5)$$

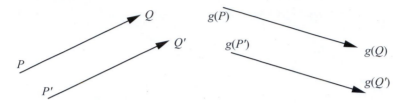

Figure 6.4b   Action of Aff(n) on vectors of $\mathbb{A}^n$.

Multiplying this out gives simply $\mathbf{y} + A\mathbf{b}$. That is, if T is the translation by $\mathbf{b}$ then gTg^{-1} is the translation by $A\mathbf{b}$.

Remark   It is easy to argue that we can write $A\mathbf{b} = g(\mathbf{b})$. In fact g acts on points of $\mathbb{A}^n$, so it also acts on based vectors $\overrightarrow{PQ}$; if $\mathbf{b} = \overrightarrow{PQ}$ then $A\mathbf{b} = \overrightarrow{g(P)g(Q)}$ (see Figure 6.4b). With this convention, we can state the conclusion in the form $g(\text{Transl}(\mathbf{b}))g^{-1} = \text{Transl}(g(\mathbf{b}))$.

I summarise the discussion of this section with the following principle, which is extremely general in scope.

Principle   *Let X be a set and $g, T : X \to X$ transformations of X. Suppose that T has some properties (or is determined by some properties) expressed in terms of data from X.*
Then the conjugate transformation $gTg^{-1} : X \to X$ has, or is determined by, the same properties expressed in terms of g applied to the same data.

Thus T fixes P gives that gTg^{-1} fixes $g(P)$, and $T = \text{Rot}(P, \theta)$ gives $gTg^{-1} = \text{Rot}(g(P), g(\theta))$.

6.5   Applications of conjugacy

6.5.1
Normal
forms

A standard 'softening up' before attacking any kind of geometric object is first to make it as simple as possible by a good choice of coordinates. We have already seen this several times in Chapter 1. For example, in 1.14 I expressed any rotation or glide of the Euclidean plane $\mathbb{E}^2$ in the form

$$\begin{pmatrix} x \\ y \end{pmatrix} \mapsto \begin{pmatrix} \cos\theta & -\sin\theta \\ \sin\theta & \cos\theta \end{pmatrix} \begin{pmatrix} x \\ y \end{pmatrix} \qquad \text{or} \qquad \begin{pmatrix} x \\ y \end{pmatrix} \mapsto \begin{pmatrix} x + a \\ -y \end{pmatrix} \tag{6}$$

with respect to a suitable Euclidean coordinate system. For the glide, you just choose coordinates so that the reflection line is the x-axis. Here the object under study is a Euclidean motion $T \in \text{Eucl}(2)$, the change of Euclidean coordinates is also an element $g \in \text{Eucl}(2)$ by the discussion in 1.12, and Theorem 1.14 says that gTg^{-1} equals one of the normal forms (6).

Similar remarks apply to Theorem 1.11. Let $T : \mathbb{R}^n \to \mathbb{R}^n$ be the orthogonal transformation of $\mathbb{R}^n$ under study. The result is that in a suitable orthonormal basis, T takes the block diagonal form of Theorem 1.11. Now $T \in O(n)$, and the change of basis is also given by an orthogonal matrix $A \in O(n)$ (because it expresses the standard basis $\{e_1, \ldots, e_n\}$ of $\mathbb{R}^n$ in terms of the special basis of Theorem 1.11, and both bases are orthogonal). Thus another way of stating the result is that ATA^{-1} equals the block diagonal matrix of Theorem 1.11.

The Jordan normal form of a matrix should be viewed as another example of conjugation. Consider any linear map $\theta : V \to V$ of an n-dimensional complex vector space V. After a choice of basis, the map θ is represented by a matrix $T \in M_{n \times n}(\mathbb{C})$. The theorem is that in a suitable basis, θ has the diagonal block form

$$
\tilde{T} = \begin{pmatrix} T_1 & & & \\ & T_2 & & \\ & & \ddots & \\ & & & T_k \end{pmatrix} \quad \text{with} \quad T_i = \begin{pmatrix} \lambda_i & 1 & & & \\ & \lambda_i & 1 & & \\ & & \ddots & & \\ & & & \lambda_i & 1 \\ & & & & \lambda_i \end{pmatrix}. \tag{7}
$$

Recall where this form comes from: the original aim is to choose a basis of V consisting of eigenvectors, which would reduce the matrix to a diagonal matrix of eigenvalues. The Jordan normal form is the next best thing if complete diagonalisation turns out to be impossible.

A coordinate change in $\mathbb{C}^n$ changes T into ATA^{-1}, where $A \in GL(n)$ expresses the change of basis; remember that separate coordinate changes in the domain and target are not allowed, because they are both the same vector space V. Hence the theorem on Jordan normal form states that if T is any matrix, for suitable choice of A the matrix ATA^{-1} has the shape of (7). If we restrict to a nonsingular matrix $T \in GL(n, \mathbb{C})$, then $T \mapsto ATA^{-1}$ is just conjugacy in $GL(n, \mathbb{C})$.

As a final example, consider permutations $T \in S_n$ of $\{1, \ldots, n\}$. Write T as

$$
t = (a_1 a_2 \cdots a_k)(a_{k+1} a_{k+2} \cdots a_{k+l}) \ldots
$$

(recall this means that under T, $(a_1 \mapsto a_2 \mapsto \cdots \mapsto a_k \mapsto a_1)$ and so on). If g is the permutation $a_i \mapsto i$ then

$$
gTg^{-1} = (12 \ldots k)(k+1 \ldots k+l) \cdots .
$$

Hence writing a permutation as a product of disjoint cycles can be thought of as describing conjugacy in the group S_n.

Remark In all the examples discussed here, finding a normal form of a transformation $T \in G$ is almost the same thing as listing the elements of G modulo the equivalence relation $T \sim gTg^{-1}$. In group theory, the equivalence classes are called *conjugacy classes* of G. For example, the above argument gives that the conjugacy classes of $GL(n, \mathbb{C})$ are exactly the Jordan normal forms (with all $\lambda_i \neq 0$). The set of conjugacy classes of a group G is one of the main protagonists in the representation theory of G.

It happens in lots of problems that we have a subset Σ of elements of a group G, and we want to know what subgroup $\langle \Sigma \rangle \subset G$ they generate. I give two quite amusing examples.

Example 1. How to walk a wardrobe   The problem of Exercise 2.12 was to prove that rotations about any two points $P \neq Q$ of $\mathbb{E}^2$ generate all direct motions of Eucl(2). I give here a solution based on conjugacy.

How to prove that I can get all the translations? First, I certainly get some translations, since the composite $\mathrm{Rot}(P, \theta) \circ \mathrm{Rot}(Q, -\theta)$ is a translation in a vector b_θ. The length of b_θ is a continuous function of θ, and is sometimes nonzero (for example, $b_{90°}$ has length $\sqrt{2}d(P, Q)$). It follows by the intermediate value theorem that we can get a translation by a vector of any fairly short length.

Now I use conjugacy: let $T = \mathrm{Transl}(b_\theta)$ be a translation, and $g = \mathrm{Rot}(P, \psi)$ a rotation. Then the conjugate gTg^{-1} is a translation by the vector $g(b_\theta)$ (see 6.4 Example 4):

$$gTg^{-1} = \mathrm{Transl}\big(g(b_\theta)\big).$$

Thus I can get a translation by any fairly short vector in any direction as a composite of my generators.

Example 2. The 15-puzzle   You can buy this puzzle in toy shops, and I am sure you all know it:

HOURS OF FUN			
1	2	3	4
5	6	7	8
9	10	11	12
13	14	15	

A legal move is to slide the blocks, restoring the blank to the bottom right. As a result of a legal move you permute the 15 numbered squares, so that clearly

$$G = \big\{\text{legal moves}\big\} \subset S_{15}.$$

Proposition   *G is the alternating group $G = A_{15}$.*

Proof. Step 1   There exists a 3-cycle $T = (11, 12, 15)$. Just rotate the three blocks in the bottom right corner.

Step 2   For any three distinct elements $a, b, c \in \{1, \dots, 15\}$, there exists a legal move g taking $11 \mapsto a$, $12 \mapsto b$, $15 \mapsto c$ (moving the other blocks any-old-how). I omit the proof, which is not hard: if you have played with the puzzle, you know from experience that you can put any 6 or 7 of the blocks anywhere you like.

Step 3   The point of this discussion: by Principle 6.4, gTg^{-1} is the 3-cycle (abc). This is easy, please think it through: $a \mapsto 11 \mapsto 12 \mapsto b, \dots$.

End of proof   For any n, the alternating group A_n is obviously generated by all 3-cycles, so that I have proved $G \supset A_{15}$. Finally, $G \subset A_{15}$. Indeed, writing 16 for the blank tile, and removing the restriction that it is always restored to the bottom right allows us to view G as a subgroup of S_{16}. But in S_{16}, every element of G is a composite of transpositions (AB) where A is the current position of the blank tile, and you must have evenly many to restore the blank to the bottom right.   QED

Note that the Proposition does not immediately explain how to solve the puzzle: knowing a group up to isomorphism does not tell you how to express its elements as words in a given set of generators.

6.5.3
The
algebraic
structure
of trans-
formation
groups

The group $\mathrm{Aff}(n)$ has two distinguished subgroups:

1. the translation subgroup $\mathbf{x} \mapsto \mathbf{x} + \mathbf{b}$, isomorphic to $\mathbb{R}^n$; and
2. the subgroup $\mathrm{GL}(n)_0$ of linear maps $\mathbf{x} \mapsto A\mathbf{x}$, isomorphic to $\mathrm{GL}(n)$ (here linear means homogeneous linear, that is, fixing 0).

Every element of $g \in \mathrm{Aff}(n)$ can be written in a unique way in the form $g \colon \mathbf{x} \mapsto A\mathbf{x} + \mathbf{b}$, that is, $g = T_\mathbf{b} \circ m_A$, where m_A is multiplication by A, and $T_\mathbf{b}$ is translation by $\mathbf{b}$. I write $g = (A, \mathbf{b})$ for short. It follows that

$$\mathrm{Aff}(n) = \mathrm{GL}(n) \times \mathbb{R}^n \quad \text{(direct product of sets)}. \tag{8}$$

However, (8) is *definitely not a direct product of groups*, because the group law is not just term by term composition: as we saw in 2.2, the composite $g_2 \circ g_1$ of $g_2 = (A_2, \mathbf{b}_2)$ and $g_1 = (A_1, \mathbf{b}_1)$ is calculated as follows:

$$\mathbf{x} \mapsto A_1\mathbf{x} + \mathbf{b}_1 \mapsto A_2(A_1\mathbf{x} + \mathbf{b}_1) + \mathbf{b}_2 = (A_2 A_1)\mathbf{x} + (\mathbf{b}_2 + A_2\mathbf{b}_1), \tag{9}$$

so that the group law is

$$(A_2, \mathbf{b}_2) \circ (A_1, \mathbf{b}_1) = (A_2 A_1, \mathbf{b}_2 + A_2\mathbf{b}_1). \tag{10}$$

This is a bit like a direct product, but the first factor A_2 interferes with the second factor $\mathbf{b}_1$ before the second factors combine.

I summarise the properties of the group given by the product (8) with the group law (10). Recall first that a *normal subgroup* of a group G is a subgroup $H \lhd G$ which is taken to itself by conjugacy in G; that is, $gHg^{-1} = H$ for all $g \in G$.

Proposition   *This setup has the following properties.*

(i)   *The translation subgroup $\mathbb{R}^n \subset \mathrm{Aff}(n)$ is a normal subgroup.*

(ii) $\mathrm{GL}(n)_0 = \{(A, 0) \mid A \in \mathrm{GL}(n)\}$ *is a subgroup of* $\mathrm{Aff}(n)$, *and is not normal.*

(iii) *The first projection $(A, \mathbf{b}) \mapsto A$ of the direct product of sets (8) defines a surjective group homomorphism* $\mathrm{Aff}(n) \to \mathrm{GL}(n)$, *under which the subgroup* $\mathrm{GL}(n)_0$ *maps isomorphically to* $\mathrm{GL}(n)$.

(iv)   *The kernel of* $\mathrm{Aff}(n) \to \mathrm{GL}(n)$ *is $\mathbb{R}^n$.*

(v)   *The action of $\mathrm{GL}(n)$ on $\mathbb{R}^n$ can be described as conjugacy in $\mathrm{Aff}(n)$.*

The *dramatis personae* of the proposition are summarised in the diagram:

$$
\begin{array}{ccc}
\mathbb{R}^n \ \lhd \ \mathrm{Aff}(n) & \to & \mathrm{GL}(n) \\
\cup & \nearrow \cong & \\
\mathrm{GL}(n)_0 & &
\end{array}
\tag{11}
$$

Proof   (i) follows from the discussion in 6.4 Example 4: the conjugate of a translation by a vector $\mathbf{b}$ is another translation, by the vector $g(\mathbf{b})$. (ii) is the same argument, although the conclusion is different: $\mathrm{GL}(n)_0$ preserves $0 \in \mathbb{R}^n$; therefore by Principle 6.4, the conjugate subgroup $g\,\mathrm{GL}(n)_0 g^{-1}$ preserves $g(0)$. Now in general $g(0) \neq 0$, and therefore $g\,\mathrm{GL}(n)_0 g^{-1} \neq \mathrm{GL}(n)_0$, so that it is not a normal subgroup.

(iii) and (iv) are obvious from the group law. For (v), note that as discussed in the remark in 6.4 Example 4, the affine group $\mathrm{Aff}(n)$ acts on $\mathbb{A}^n$, and also acts on vectors of $\mathbb{A}^n$, taking $\overrightarrow{PQ}$ to $\overrightarrow{g(P)g(Q)}$. This gives a well defined action of $\mathrm{Aff}(n)$ on $\mathbb{R}^n$: indeed $\overrightarrow{PQ} = \overrightarrow{P'Q'}$ means that $PQQ'P'$ is a parallelogram; an affine map takes a parallelogram into another parallelogram, so that also $\overrightarrow{g(P)g(Q)} = \overrightarrow{g(P')g(Q')}$ (compare Figure 6.4b). Thus the projection $(A, \mathbf{b}) \mapsto A$ is just the action of $\mathrm{Aff}(n)$ on $\mathbb{R}^n$ (thought of as the free vectors of $\mathbb{A}^n$). But this is also the action of $\mathrm{Aff}(n)$ by conjugacy on translations by vectors in $\mathbb{R}^n$.   QED

Remarks

1. The same holds for the Euclidean group, with $\mathrm{O}(n)$ in place of $\mathrm{GL}(n)$. That is, the same scenario can be replayed word for word with the new cast of players:

$$
\begin{array}{ccc}
\mathbb{R}^n \ \lhd \ \mathrm{Eucl}(n) & \to & \mathrm{O}(n) \\
\cup & \nearrow \cong & \\
\mathrm{O}(n)_0 & &
\end{array}
\tag{12}
$$

2. Philosophy: the groups are contained in the geometry, as transformation groups. However, the geometry is also contained in the algebra: the vector space $\mathbb{R}^n$ and the action of $GL(n)$ on it are contained in the group structure of $Aff(n)$. To spell this out, $\mathbb{R}^n$ is the subgroup of translations in $Aff(n)$, and the action of $GL(n)$ on $\mathbb{R}^n$ is the conjugacy action of $Aff(n)$ on the translations.

The affine space $\mathbb{A}^n$ and the action of $Aff(n)$ on it are also buried in the group structure of $Aff(n)$. Indeed, $GL(n)_0$ is the subgroup of elements preserving 0, and its conjugates are the subgroups $GL(n)_P$ preserving other points $P \in \mathbb{A}^n$. Thus $\mathbb{A}^n$ is in one-to-one correspondence with these conjugates.

3. This remark is intended for students who know about abstract groups, and what it means for an abstract group to *act on* a mathematical structure. (Some details of what is involved are discussed in Exercise 6.17; see also Section 9.2.) There is a general notion of *semidirect product* $G \ltimes H$ of abstract groups: if a group G *acts on* a group H *by group homomorphisms*, then $G \ltimes H$ is the set of pairs (A, b) with $A \in G$ and $b \in H$ with the group law $(A_2, b_2) \circ (A_1, b_1) = (A_2 A_1, b_2(A_2 b_1))$. It is an easy exercise in abstract groups (Exercise 6.17) to see that this makes $G \ltimes H$ into a group, which fits into a diagram like (11).

6.6 Discrete reflection groups

Recall from 2.6 that reflections generate all motions of Euclidean space. In general, a group generated by some set of reflections of $\mathbb{E}^n$ is called a *reflection group*. Of special interest are relatively 'small' reflection groups; in Example 1, the group is finite; in Examples 2–3 it is infinite but 'discrete' that is, group elements are in a sense 'well spaced'. I do not have space here to elaborate on the theory but I give the most basic examples.

Example 1. Kaleidoscope   Two planar reflections in Euclidean lines Π_1, Π_2 meeting at an angle $\theta = \pi/n$ generate a finite group (Figure 6.6a). If s_1 and s_2 are the two reflections then $s_2 \circ s_1$ is a rotation through $2\pi/n$, so $(s_2 \circ s_1)^n = id$. As an abstract group this is the dihedral group D_{2n}, containing the cyclic group generated by the rotation $s_2 \circ s_1$ as a subgroup of index 2; see Exercise 6.5 for details.

By contrast, to get an idea of what I mean by 'well spaced' group elements, think of the group generated by reflections in two lines that meet at an angle that is an irrational multiple of π.

Example 2. Barber's shop   Reflections in two parallel mirrors Π_1, Π_2. This is the infinite dihedral group $D_{2\infty}$ generated by s_1 and s_2 with $s_1^2 = s_2^2 = id$, and no other relations. It contains the infinite cyclic group generated by the translation $s_1 \circ s_2$ as a subgroup of order 2.

Example 3. Musée Grévin   The Musée Grévin is the Paris equivalent of Madame Tussaud's (the waxworks). They have a spectacular show in which members of the

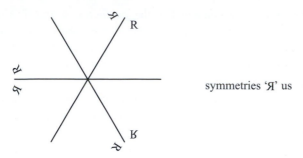

symmetries 'Я' us

Figure 6.6a   Kaleidoscope.

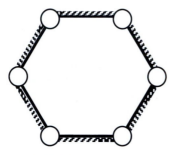

Figure 6.6b   'Musée Grévin'.

paying public and their children stand inside a kaleidoscope made of mirrors forming a regular hexagon. At the angles of the hexagon they put exotically decorated columns (Figure 6.6b). When the lights come on, you have the impression of standing in an infinite honeycomb pattern containing periodically arranged family groups with babies in pushchairs. The reflection group here is the group generated by reflections in the six sides of the hexagon. See Exercise 6.6 for details.

Reflection groups turn up all over the place in mathematics, from the theory of Platonic solids through the theory of crystals, Coxeter groups, Lie theory (the Weyl group), to Riemann surfaces, which are related to Fuchsian groups acting on hyperbolic rather than Euclidean space. For a first port of call, consult Coxeter [5].

Exercises

6.1   Prove that $(n + 1) \times (n + 1)$ matrixes with the block form $\left(\begin{smallmatrix} A & b \\ 0 & 1 \end{smallmatrix} \right)$ where A is $n \times n$ and b is $n \times 1$ form a group isomorphic to Aff(n). Verify Proposition 6.5.3 in these terms.

6.2   A similarity $s \colon \mathbb{E}^n \to \mathbb{E}^n$ is a transformation which scales distances by a constant factor $\lambda > 0$ (that is, $d(s(x), s(y)) = \lambda d(x, y)$ for all x, y). Here λ depends on s only.
(a) Prove that the set of similarities is a transformation group Sim(n) of $\mathbb{E}^n$.
(b) Sim(n) does not preserve distances in $\mathbb{E}^n$. Prove that it preserves angles.

(c) Show how to use the scaling factor λ to define a group homomorphism $\mathrm{Sim}(n) \twoheadrightarrow \mathbb{R}_{>0}$ with $\mathrm{Eucl}(n)$ as its kernel.

6.3 Prove that the diagonal scalar matrixes $\mathrm{diag}(\lambda, \lambda, \ldots, \lambda)$ form a subgroup of $\mathrm{GL}(n)$, equal to the centre ($=$ the set of elements that commute with every matrix). Prove that $\mathrm{PGL}(n + 1)$ is the quotient of $\mathrm{GL}(n + 1)$ by its centre (compare 5.5).

6.4 Let G be a finite group of motions of $\mathbb{E}^2$. Prove that there is a point of $\mathbb{E}^2$ fixed by every element of G. [Hint: take the average.] Deduce a description of every element of $\mathrm{Eucl}(n)$ of finite order.

6.5 Let S_n be the regular n-gon in $\mathbb{E}^2$, for $n \geq 3$, and let D_{2n} be the symmetry group of S_n. Show that

(a) every element of D_{2n} fixes the centre of S;

(b) D_{2n} contains n rotations (including the identity), which form a subgroup H_n of D_{2n} isomorphic to the cyclic group of order n;

(c) D_{2n} also contains n reflections, and no further elements, hence has order $2n$;

(d) D_{2n} is isomorphic to the reflection group of 6.6 Example 1.

Denoting by a one of the reflections and by b a rotation by a smallest angle, write out the group elements in terms of a, b. Find the relations holding between a and b. Deduce from your relations that H_n is a normal subgroup of D_{2n}. [Hint: if you get stuck, first do the case of the square $\square$ in $\mathbb{E}^2$ with vertexes $(\pm1, \pm1)$; here it is easy to write out the elements of D_8 as a set of matrixes, and doing this case gives you all the psychological support needed to do the general case.]

The group D_{2n} is called the *dihedral group of order* $2n$, a group which occurs in many guises in and out of geometry.

6.6 The reflection group G corresponding to the *Musée Grévin* described in 6.6 Example 3 and Figure 6.6b is the group generated by reflections in the sides of a regular hexagon H, which acts on $\mathbb{E}^2$ preserving the honeycomb tiling by regular hexagons. Show that

(a) G contains the reflections in the 3 diagonals of H, generating a group of symmetries of H isomorphic to S_3.

(b) Translations in G form a normal subgroup $\mathbb{Z} \oplus \mathbb{Z} \cong T \triangleleft G$, with quotient $G/T \cong S_3$.

(c) G is of index 2 in the full group of symmetries of the hexagonal tiling. [Hint: colour vertexes of the honeycomb tiling alternately black and white.]

Exercises in conjugacy.

6.7 Write $\mathrm{Stab}_G(x) \subset G$ for the set of elements of G that fix x (the *stabiliser* of x in G); prove that $\mathrm{Stab}_G(x)$ is a subgroup.

Let $G \subset \mathrm{Trans}\, X$ be a transformation group of a set X. For $x \in X$ and $g, t \in G$, prove that t fixes x if and only if gtg^{-1} fixes $g(x)$ (compare 6.4 (3)).

Deduce that $\mathrm{Stab}_G(gx)$ is the conjugate subgroup $g\, \mathrm{Stab}_G(x)g^{-1}$.

6.8 Prove that the distinction between direct and opposite motion (Definition 1.10) is independent of the choice of coordinates. [Hint: let T be the motion in question, and $g \in \mathrm{Eucl}(n)$ a coordinate change. By the principle of 6.4, T is expressed in the new coordinates by gTg^{-1}. It remains to calculate the linear part of gTg^{-1} and its determinant.]

6.9 G is a group. Prove that conjugacy is an equivalence relation on G. That is, the relation $g \sim g'$ if and only if g and g' are conjugate in G is an equivalence relation. Determine all the conjugacy classes in the symmetric group S_4.

6.10   Prove that any two translations Transl($\mathbf{b}$) by a nonzero vector $\mathbf{b}$ are conjugate in Aff(2). (Compare 6.4 Example 4.) Which translations in Eucl(2) are conjugate?

6.11   Prove that two rotations of $\mathbb{E}^2$ are conjugate in Eucl(2) if and only if the absolute value of the angles are equal.

6.12   Use Principle 6.4 and Theorem 1.14 to list the conjugacy classes of Eucl(2). [Hint: every motion is conjugate to a standard type. You have to say when two standard types are conjugate, and to choose exactly one normal form from each conjugacy class.]

6.13   Consider the field $\mathbb{F}_p = \mathbb{Z}/p$ with p elements. The projective line $\mathbb{P}^1_{\mathbb{F}_p}$ over $\mathbb{F}_p$ is the set of 1-dimensional vector subspaces of $\mathbb{F}_p^2$, or equivalently, the set $(\mathbb{F}_p^2 \setminus 0)/\sim$. It has $p + 1$ elements, called $0, 1, 2, \ldots, p - 1, \infty$. Use Theorem 5.5 to prove that the general linear group PGL$(2, \mathbb{F}_p)$ has order $(p + 1) \cdot p \cdot (p - 1)$.

6.14   Specialise to $p = 5$, and the action of PGL$(2, \mathbb{F}_5)$ on the 6 points $\{0, 1, 2, 3, 4, \infty\}$ of $\mathbb{P}^1_{\mathbb{F}_5}$. Write down the 3 maps

$$x \mapsto x + 1, \quad x \mapsto 2x \quad \text{and} \quad x \mapsto 2 - 2/x$$

(where x is an affine coordinate) as permutations of these 6 elements.

6.15   Determine the subgroup of S_6 (the symmetric group on 6 elements) generated by the 2 elements $\sigma = (abcd)$ and $\tau = (cdef)$. [Hint: if you play around for a while with lots of combinations of the generators, you will notice that it is 3-transitive, but you only get a few cycle types, so it is probably quite a bit smaller than the whole of S_6.]

6.16   (Harder) Determine the subgroup G of the symmetric group S_7 generated by $\sigma = (1234)$ and $\tau = (34567)$. [Hint: the answer is S_7. Indeed, G is obviously 3 or 4-transitive: as with the 15 puzzle (6.5.2 Example 2), you can put any 3 elements anywhere you like by messing around with the given generators. G also contains an odd permutation σ, so is not contained in the alternating group A_7. To complete the proof, you need to find a transposition or a 3-cycle; then G must contain A_7 by the same principle as 6.5.2 Example 2.]

6.17   (Assumes abstract group theory) Let G and H be abstract groups. Say what it means for G to act on H by group homomorphisms $(A, b) \mapsto Ab$. Under this assumption, prove that the multiplication

$$(A_2, b_2) \circ (A_1, b_1) = (A_2 A_1, b_2(A_2 b_1)) \quad \text{for } A_i \in G \text{ and } b_i \in H$$

makes the direct product $G \times H$ into an abstract group $G \ltimes H$, such that the assertions of Proposition 6.5.3 hold for it.

7 Topology

The word topology in the context of this course has two quite different meanings:

'Point-set topology'   Slogan: a topological space is a 'metric space without a metric'. In analysis, this idea leads to a fairly minor generalisation of the definition of metric space, but the definition of topology has applications in other areas of math, where it turns out to be logical or algebraic in content. I give the abstract definition and some examples of topological spaces that are definitely not metric. This is an important ingredient in all advanced math (algebra, analysis, arithmetic, geometry, logic, etc.). Topology has lots of advantages even when the only spaces of interest are metric spaces. It provides, in particular, a simple rigorous language for 'sufficiently near' without epsilons and deltas.

'Rubber-sheet geometry'   The abstract language gives us tools to study spaces that are geometric in origin, such as the torus and the Möbius strip. Geometric concepts in topology include the winding number and the number of holes of a surface.

Here is a sample of the results proved in this chapter.

1. If $f\colon S^1 \to \Sigma \subset \mathbb{R}^2$ is bijective and continuous, then the inverse map $f^{-1}\colon \Sigma \to S^1$ is also continuous; that is, f is a *homeomorphism*. Joke: topology is geometry in which

$$\heartsuit = 0.$$

Imagine trying to prove this from first principles! The point is that f can be very complicated, and f^{-1} might not be given by any simple function.
2. The cylinder is different from the Möbius strip.
3. The winding number: let $\varphi\colon [0, 1] \to \mathbb{R}^2 \setminus (0, 0)$ be a continuous map with $\varphi(0) = \varphi(1)$. Then the number of times φ winds around the origin is not changed by deforming the loop continuously; in other words, the winding number is a *homotopy invariant* of the map φ.

7.1 Definition of a topological space

Let X be a set. A *topology* on X is a collection $\mathcal{T}$ of subsets of X satisfying the following three axioms:

- **finite intersection** $U_1, \ldots, U_n \in \mathcal{T} \implies U_1 \cap \cdots \cap U_n \in \mathcal{T}$;
- **arbitrary union** $U_\lambda \in \mathcal{T}$ for $\lambda \in \Lambda \implies \bigcup_{\lambda \in \Lambda} U_\lambda \in \mathcal{T}$, where Λ is an arbitrary indexing set;
- **conventions on empty set** $\emptyset, X \in \mathcal{T}$.

A *topological space* is a pair $X, \mathcal{T}$ consisting of a set X and a topology $\mathcal{T} = \mathcal{T}_X$ on it. $U \in \mathcal{T}$ is called an *open set* of the topology $\mathcal{T}$. We often speak of the topological space X and its open sets U, omitting $\mathcal{T}$ from the notation when it is clear what topology is intended. $V \subset X$ is *closed* if its complement is open; the topology could be specified equally well by the collection of closed sets, which enjoys finite union and arbitrary intersection. If $Z \subset X$, the *closure* of Z, denoted $\overline{Z}$, is the intersection of all closed sets containing Z. By the arbitrary intersection property of closed sets, $\overline{Z}$ is closed; it clearly contains Z. A *neighbourhood* of a point $x \in X$ is any subset $V \subset X$ containing an open set containing x.

We will see presently that if X is a metric space then there is a natural choice of open sets of X which form a topology. Here are some simpler examples.

Example 1   Let $X = \{P_1, P_2, P_3\}$ be a set consisting of 3 points, and

$$\mathcal{T}_X = \{\emptyset, \{P_1\}, \{P_1, P_2\}, X\}.$$

Then $\{P_1\}$ is open, but every neighbourhood of P_2 contains P_1, and every neighbourhood of P_3 contains both P_1, P_2.

Example 2   There are two extreme topologies defined on any set X. The *discrete topology* has every subset open. The *indiscrete topology* has no open sets except $\emptyset$ and X itself.

Example 3   The *cofinite topology* on an infinite set X is the topology for which the open sets are $\emptyset$ or the complements of finite sets; that is, $U \subset X$ is open if and only if either $U = \emptyset$ or $X \setminus U$ is finite; it is obvious that this satisfies finite intersection and arbitrary union. In this topology, if $x \in U$ and $y \in V$ are neighbourhoods of any two points then $U \cap V$ is also the complement of a finite set, and hence nonempty.

7.2 Motivation from metric spaces

Let (X, d) and (Y, d') be metric spaces (see Appendix A if you need reminding what this means) and $f : X \to Y$ a map. By definition, f is *continuous* if for every $x \in X$ and for any given $\varepsilon > 0$, there exists $\delta > 0$ such that $d(x, y) < \delta \implies d'(f(x), f(y)) < \varepsilon$.

The intuitive meaning is clear without epsilons and deltas: if $x \in X$ is any given point, I can guarantee that $f(y)$ is arbitrarily close to $f(x)$ by forcing y to be sufficiently close to x.

The idea of topology on a space is to break up the definition of continuity into two steps. First use the metric to derive the open sets and neighbourhoods of points; then describe continuity in terms of open sets.

Definition   If (X, d) is a metric space, a set $U \subset X$ is a *neighbourhood* of x if $B(x, \varepsilon) \subset U$ for some ε. Here $B(x, \varepsilon)$ is the open ball of radius ε centred at x; if you cannot guess the formal definition, look in Appendix A. A set $U \subset X$ is *open* if it is a neighbourhood of every one of its points, that is, for all $x \in U$, $B(x, \varepsilon) \subset U$ for some ε. The open sets U of X form a topology on X, the *metric topology* of (X, d). (See Exercise 7.1.)

Equivalent conditions   Standard easy result on metric spaces:

$$f \text{ is continuous} \iff \begin{array}{l} \forall\, x \in X \text{ and } \forall \text{ neighbourhood } V \subset Y, \\ f^{-1}V \subset X \text{ is a neighbourhood of } x \end{array}$$

$$\iff \forall \text{ open } V \subset Y, f^{-1}V \subset X \text{ is open.}$$

In other words, the 'epsilon-delta' definition of continuity for metric spaces can be replaced by an equivalent condition which involves only open sets of the metric topology. I will adopt this equivalent condition in 7.3 to define continuity for a map between arbitrary abstract topological spaces.

The idea of a topological space is a natural abstraction and generalisation of the idea of a metric space. When going from a metric space (X, d) to the corresponding topological space, we forget the metric, and keep only the notion of neighbourhoods, or equivalently open sets. There are several advantages. In the context of metric spaces, closeness means that the distance $d(x, y)$ is small. But just as some things in life have a value that cannot be expressed as a sum of money, in some contexts closeness cannot always be expressed as a distance measured as a real number. In particular, the following three properties are forced on metric spaces by definition, but are optional for topological spaces.

1. Symmetry: in a metric space, x is close to y if and only if y is close to x.
2. Hausdorff property: given two points $x \neq y \in X$, there exist disjoint open sets $x \in U$ and $y \in V$ (see Figure 7.2a).
3. Countable neighbourhoods: given a point $x \in X$ of a metric space, consider the family $B_n = B(x, \frac{1}{n})$. Then B_n are neighbourhoods of x; they are countable in number; every

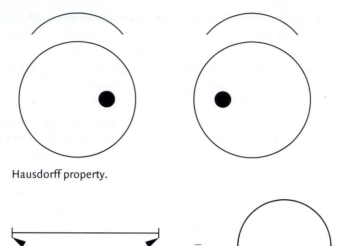

Figure 7.2a Hausdorff property.

Figure 7.2b $S^1 = [0, 1]$ with the ends identified.

neighbourhood of x contains a B_n; and $\bigcap B_n = \{x\}$. This can be used in convergence arguments in analysis (see Exercise 7.4).

The idea of having the open sets specified as the basic construction is of course more abstract and less intuitive than definitions in first analysis or metric spaces courses, but abstractness has its own advantages. In many cases, the spaces I am interested in may actually be metric spaces, but I may not really care about the distances, just in what it means for $d(x, y) \ll 1$. For example, if you think of the circle $S^1 \subset \mathbb{R}^2$ as the identification space obtained by glueing together the ends of the interval $[0, 1]$, then S^1 is a metric space, with metric

$$d_{S^1}(P, Q) = \min \left(\begin{matrix} d_{[0,1]}(P, Q), & d_{[0,1]}(0, P) + d_{[0,1]}(Q, 1), \\ d_{[0,1]}(0, Q) + d_{[0,1]}(P, 1) \end{matrix} \right),$$

which is a fairly tedious expression to work with; but I really do not care about the metric, only the system of arbitrarily small neighbourhoods of points. A small neighbourhood of any point other than the 'seam' P_0, the image of the endpoints $0, 1$, is given by $(x - \varepsilon, x + \varepsilon)$ from the interior of the interval. For P_0, you glue together small neighbourhoods of the glued endpoints: $[0, \varepsilon) \cup (1 - \varepsilon, 1]$; see Figure 7.2b.

As a final example, note that the discrete topology on any set X, defined in 7.1 Example 2, is metric: just set $d(x, y) = 1$ for every $x \neq y$. On the other hand, the indiscrete topology is not metric.

7.3 Continuous maps and homeomorphisms

7.3.1 Definition of a continuous map

If X and Y are topological spaces, a map $f : X \to Y$ is *continuous* if

$$f^{-1}(U) \subset X \text{ is open for every open } U \subset Y.$$

Notice that I am already omitting mention of the topologies $\mathcal{T}_X$ and $\mathcal{T}_Y$. To use the language literally, I should have said the following: let $X, \mathcal{T}_X$ and $Y, \mathcal{T}_Y$ be topological spaces, then f is continuous if

$$U \in \mathcal{T}_Y \implies f^{-1}(U) \in \mathcal{T}_X.$$

Example 1 If X is any set with the discrete topology of 7.1 Example 2, then every map $X \to Y$ from X to any topological space is continuous. If X has the indiscrete topology, then every map $Y \to X$ from any topological space to X is continuous.

Example 2 Consider an infinite field k with the cofinite topology on it (see 7.1 Example 3). Let $f : k \to k$ be a polynomial map given by $a \mapsto f(a)$, where f is a polynomial in one variable. Then f is continuous. For $U \subset k$ is open if and only if $U = \emptyset$ or U is the complement of a finite set, say $U = k \setminus \{b_1, \dots, b_n\}$; then $f(x) = b_i$ has at most $\deg f$ solutions, so that $f^{-1}(U)$ is also the complement of a finite set.

7.3.2 Definition of a homeomorphism

A map $f : X \to Y$ is a *homeomorphism* if f is bijective, and both f and f^{-1} are continuous. This means that

$$f : X \leftrightarrow Y \qquad \text{and} \qquad \mathcal{T}_X \leftrightarrow \mathcal{T}_Y,$$

or in other words, f is an isomorphism of all the structure there is. X and Y are homeomorphic, written $X \simeq Y$, if there exists a homeomorphism $f : X \to Y$.

Example 3 An open interval (a, b) is homeomorphic to the real line, $(a, b) \simeq \mathbb{R}$. For example, the map

$$f : (0, 1) \to \mathbb{R} \quad \text{defined by} \quad f(x) = \frac{-1}{x} + \frac{1}{1-x} = \frac{2x - 1}{x(1-x)}$$

is a homeomorphism, illustrated in Figure 7.3a.

Example 4 The square is homeomorphic to the circle in $\mathbb{R}^2$. To see this, put the square inside the circle and project out from an interior point (see Figure 7.3b). A similar radial projection argument shows also that the full square is homeomorphic to the closed disc $\{x^2 + y^2 \le 1\} \subset \mathbb{R}^2$. In Theorem 7.14 below I show that if $f : S^1 \to \mathbb{R}^2$ is any one-to-one and continuous map (that is, a simple closed curve) then f is a homeomorphism.

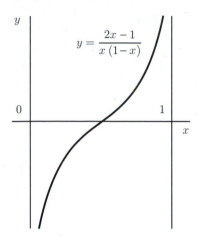

Figure 7.3a $(0, 1) \simeq \mathbb{R}$.

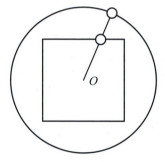

Figure 7.3b   Squaring the circle.

Example 5   If (X, d) and (Y, d') are metric spaces and $f : X \to Y$ is an isometry, then f is a homeomorphism. Note however a map f can set up a homeomorphism between (the metric topologies of) metric spaces *without* being an isometry, as in Examples 3 and 4 above. Being homeomorphic is a much coarser relation on metric spaces than being isometric. Example 7.4.2 discusses this issue from a slightly different point of view.

7.3.3
Homeomor-
phisms and
the Erlangen
program

The group $\mathrm{Homeo}(X)$ of self-homeomorphisms is a transformation group of the topological space X (compare 6.1). In the framework of the Erlangen program of Section 6.3, topology can be viewed as the study of properties invariant under $\mathrm{Homeo}(X)$.

The homeomorphism group of $X = \mathbb{R}$ is already an uncomfortably large infinite group, and its action mixes up the points of $\mathbb{R}$ like anything, so at first sight it seems hard to imagine how any invariant properties can survive. However, such properties do exist; one example is *between-ness*, or *separation*, derived from the order relation of $\mathbb{R}$: a homeomorphism f takes three real numbers x, y, z with y between x and z into $f(x), f(y), f(z)$ with $f(y)$ between $f(x)$ and $f(z)$; this follows at once from the intermediate value theorem.

If a geometry has lines which are homeomorphic copies of the real line $\mathbb{R}$, then the separation property can be formulated in the geometry: a point cuts a line into two disconnected subsets, and hence it makes sense to ask whether a point Q on a line *lies between* two other points P, R. Euclidean and hyperbolic geometry are examples where this property holds. In contrast, the lines (great circles) of spherical geometry have the topology of the circle S^1, so they have the 'no separation' property: cutting a point leaves behind a set which is still connected. See 9.1 for the historic significance of this issue.

7.3.4
The homeo-
morphism
problem

The following 5 spaces are not homeomorphic (for proofs, please be patient until 7.4.4):

(1) the closed interval $[a, b]$;
(2) the open interval $(a, b) \simeq \mathbb{R}$;
(3) the circle S^1;
(4) the plane $\mathbb{R}^2$;
(5) the sphere $S^2 \subset \mathbb{R}^3$.

The examples here and in 7.3.2 illustrate an important general point. If you want to prove that two given topological spaces X and Y are homeomorphic, then it is your job to supply a homeomorphism $f \colon X \to Y$, for example by a geometric construction; or at least, to prove that one exists. On the other hand, to prove that X and Y are *not* homeomorphic, you need to find some property of spaces that is the same for homeomorphic spaces, but different for X and Y. This is called the 'homeomorphism problem'.

The next few sections introduce some basic notions of topology and use them to prove assertions of this type. Algebraic topology has as one of its main aims to develop systematic invariants of topological spaces that can be used to prove that spaces are not homeomorphic, notably the fundamental group $\pi_1(X, x_0)$ and homology groups $H_i(X, \mathbb{Z})$; but in this book I work only with very simple ideas.

7.4 Topological properties

Some properties of a topological space depend only on the topology. A *topological property* of topological spaces is a property that can be expressed in terms of points and open sets only. *Homeomorphisms preserve topological properties.*

For example, if X is a metric space, then *bounded* is not a topological property: it depends on distance ($d(x, y) \le K$ for some K), and not just on the topology. Thus $(a, b) \simeq \mathbb{R}$ (see Figure 7.3a), but the left-hand side is bounded, while the right-hand side is not.

7.4.1
Connected
space

A topological space X is *connected*, if it cannot be written as a disjoint union of two nonempty open subsets; that is, there does not exist any decomposition

$$X = U_1 \sqcup U_2 \quad \text{with } U_1, U_2 \text{ open,}$$

where $\sqcup$ denotes disjoint union.

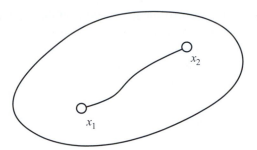

Figure 7.4a   Path connected set.

A *path* in a space X is a continuous map $\varphi\colon [0, 1] \to X$; X is *path connected* if for any two points $x_1, x_2 \in X$, there exists a path φ with $\varphi(0) = x_1$ and $\varphi(1) = x_2$ (that is, any two points can be joined by a path). (See Figure 7.4a.)

Connected and path connected are both topological properties, since only open sets and continuous maps appear in their definitions. Thus given two spaces X, Y, if $X \simeq Y$ then X and Y are either both (path) connected or both (path) disconnected.

Lemma

(1)   *The interval* $[0, 1]$ *is connected.*
(2)   *A path connected set is connected.*

Proof   For (1), suppose $[0, 1] = U_1 \sqcup U_2$ with opens $U_1, U_2 \subsetneq [0, 1]$. Say $0 \in U_1$, and consider

$$z = \sup\{x \mid [0, x] \subset U_1\},$$

where sup is least upper bound from your first analysis course. The sup exists by the completeness axiom of the reals. If $z \in U_1$, then because U_1 is open, there is a neighbourhood of z in U_1, that is, $[z, z + \varepsilon) \subset U_1$ for some $\varepsilon > 0$, so z is not an upper bound. If $z \in U_2$, there is a neighbourhood of z in U_2, so an interval $(z - \varepsilon, z]$ disjoint from U_1 and so $z - \varepsilon$ is a strictly smaller lower bound, which also contradicts the definition of z as sup. (The proof is the same as that of the intermediate value theorem in a first analysis course.)

To show (2), suppose X is path connected and $X = U_1 \sqcup U_2$ with opens $U_1, U_2 \subsetneq X$. Then choose $x \in U_1$ and $y \in U_2$ and apply the definition of path connected, so that there is a continuous map $\varphi\colon [0, 1] \to X$ with $\varphi(0) = x$ and $\varphi(1) = y$. Then $[0, 1] = \varphi^{-1}(U_1) \sqcup \varphi^{-1}(U_2)$ is a disjoint union, with both $\varphi^{-1}(U_1)$ and $\varphi^{-1}(U_2)$ open and nonempty, which contradicts (1).   QED

If X is any topological space, define a relation on X by setting $x \sim y$ if and only if there is a connected subset U of X containing x, y. It is clear that $\sim$ is symmetric and reflexive, and a bit of thought tells you that it is also transitive, hence it is an equivalence relation. Equivalence classes of $\sim$ are called *components* of the topological space X.

Remark The property used to define a path connected space corresponds to our usual perception of 'connectedness': you can get from point A to point B using an unbroken 'path'. In the context of surface travel, mainland Eurasia forms a path connected space but the United States does not: you cannot get from New York to Alaska without crossing Canada or going by air or sea. However, in the context of general topological spaces, connectedness as defined above, without reference to paths, is preferable. By the Lemma, path connectedness implies this more general form of connectedness. Similar remarks apply to components: the definition is a natural extension of the obvious notion under which the connected components of the United Kingdom include mainland Britain and mainland Northern Ireland, along with any number of smaller islands around the coast.

**7.4.2
Compact
space**

The space X is *compact* if

for every cover $X = \bigcup_{\lambda \in \Lambda} U_\lambda$ of X by an arbitrary collection of opens U_λ, there exists a finite number of indexes $\lambda_1, \ldots, \lambda_n \in \Lambda$ such that $X = \bigcup_{i=1}^{n} U_{\lambda_i}$.

(Slogan: *every open cover has a finite subcover.*) This property manifestly depends only on open sets.

A sequence of points $a_1, a_2, \ldots$ in a topological space X *converges* to a limit $l \in X$, written $a_{i_n} \to l$, if for any neighbourhood U of l, the a_i are eventually all in U. In other words,

for every open set U of X with $l \in U$, there exists n_0 such that $a_i \in U$ for all $i \geq n_0$.

In other words, $a_1, a_2, \ldots$ tend to $l \in X$ if, *for any measure of closeness*, the a_i are *eventually all close* to l.

The space X is *sequentially compact* if every sequence has a convergent subsequence, that is, for every infinite sequence $a_1, a_2, \ldots$ of points of X, there exists a point $x \in X$ and a sequence $i_1, i_2, \ldots$ of indexes such that $a_{i_n} \to x$. (Slogan: *every sequence has a convergent subsequence.*)

The following statement relates these notions to each other and to more familiar ones in *metric spaces*.

Proposition

(1) *For V a subset of $\mathbb{R}^n$ with its usual (Euclidean) metric,*

$$V \text{ is closed and bounded } \iff V \text{ is sequentially compact.}$$

(2) *For X any metric space and $V \subset X$ a subset,*

$$V \text{ is sequentially compact } \iff V \text{ is compact.}$$

Here is a brief discussion of where you can find this in the literature. Compactness is the subject of Sutherland [24], Chapter 5. The statement that a closed bounded subset of $\mathbb{R}^n$ is compact is the Heine–Borel theorem, proved in [24], Theorem 5.3.1 for $n = 1$, and in general (by reducing to the case $n = 1$) in Theorem 5.7.1. Compact implies sequentially compact (in a metric space) is proved in [24], Theorem 7.2.6. The other way round, sequentially compact implies compact (in a metric space), is proved in [24], Chapter 7. The proof is a bit tricky, but Sutherland breaks it up into 3 self-contained steps, each of which takes a half-page. (See also, for example, Rudin [21], 2.31–2.40.)

This is not primarily a course on foundational stuff in metric spaces, and I take a common sense approach: when I am working in a metric space, I use compact or sequentially compact more-or-less interchangeably. With general topological spaces, the language of compactness is more natural and more convenient.

Example   Consider the *n-sphere*

$$S^n = \{(x_1, \ldots, x_n) \in \mathbb{R}^n \mid x_1^2 + \cdots + x_n^2 = 1\}.$$

You have already seen two different metrics on S^n: one is the Euclidean distance of points on $S^n \subset \mathbb{R}^n$, and the other one is the spherical distance $d(\mathbf{x}, \mathbf{y}) = \arccos(\mathbf{x} \cdot \mathbf{y})$ (see 3.1 and compare Exercise 3.10). However, points are close to each other in one of the metrics if and only if they are close in the other; said differently, the metric topologies given by the two metrics are *the same*. Under the Euclidean metric inherited from $\mathbb{R}^n$, the set S^n is bounded (distance 1 from the origin) and closed (clearly) so S^n is compact by (1) of the Proposition.

7.4.3
Continuous image of a compact space is compact

Proposition   *Let X, Y be topological spaces and $f \colon X \to Y$ a surjective continuous map. Then if X is compact, so is Y.*

Proof   You just have to write out the definitions: if $Y = \bigcup V_\lambda$, an arbitrary union of open sets, let $U_\lambda = f^{-1}(V_\lambda)$. Then U_λ is open, and $X = \bigcup U_\lambda$. Therefore there exists a finite set of indexes $\lambda_1, \ldots, \lambda_n$ such that $X = \bigcup_{i=1}^{n} U_{\lambda_i}$. Finally,

$$Y = f(X) = \bigcup_{i=1}^{n} f(U_{\lambda_i}) = \bigcup_{i=1}^{n} V_{\lambda_i}. \quad \text{QED}$$

Pretty easy wasn't it? This shows what a convenient property compactness is. Compare the result in analysis: a continuous function $f \colon [a, b] \to \mathbb{R}$ is bounded and attains its bound. This is hard to prove from first principles, but is really easy once you have established the definition of compactness, and proved Proposition 7.4.2.

The notion of compactness is a powerful tool, and you should learn to use it, even if you put off studying the proofs until later. A typical use is the kind of 'continuity implies uniform continuity' argument used all over the place in analysis. If $f \colon [a, b] \to \mathbb{R}$ is continuous, then given $\varepsilon > 0$, for all $x \in [a, b]$, you can force $f(x')$ that close to $f(x)$ by squeezing x' within δ of x; here δ depends on x, but compactness allows you to choose one δ that works uniformly for all $x \in [0, 1]$.

There is a famous Bertrand Russell quotation about the advantages of the axiomatic method: *they are the advantages of theft over honest labour*. You must either understand a proof of the Heine–Borel theorem (e.g. Sutherland [24], Theorem 5.7.1), or take it on trust as an axiom and accept the advantages.

7·4·4
An
application
of topo-
logical
properties

The notions set up so far are already enough to give a proof of the statement in 7.3.4. For example, the topological nature of connectedness implies that $(a, b) \not\cong [a, b]$, because any point disconnects the left-hand side. In more detail, if $x \in (a, b)$ is any point then it disconnects (a, b) into two disjoint open intervals (a, x) and (x, b); if $\varphi\colon [a, b] \to (a, b)$ were a homeomorphism, then $\varphi(a) = x \in (a, b)$ would be an interior point, so $(a, b) \setminus x$ would be disconnected, whereas $[a, b] \setminus \{a\} = (a, b]$ is connected. For exactly similar reasons,

$$S^1 \not\cong [a, b], \mathbb{R}^2, S^2 \quad \text{(any 2 points disconnect the left-hand side)}$$

$$\mathbb{R} \not\cong \mathbb{R}^2 \quad \text{(any point disconnects the left-hand side)}$$

$$[a, b] \not\cong \mathbb{R}^2 \text{ or } S^2 \quad \text{(any 3 points disconnect the left-hand side)}.$$

To complete the argument, note that

$$[a, b], S^1, S^2 \not\cong (a, b), \mathbb{R}, \mathbb{R}^2;$$

because all the spaces on the left-hand side are compact, and all those on the right are not.

7·5 Subspace and quotient topology

If X is a topological space and $Z \subset X$ a subset, write $i\colon Z \hookrightarrow X$ for the inclusion map, that is $i(z) = z \in X$ for every $z \in Z$. Then the *subspace topology* of Z is the topology whose open sets are of the form $U \cap Z$, where U is an open of X. If X is a metric space with the topology defined by the metric d, then the subspace topology of Z is also metric, defined by the same metric restricted to Z. This definition of the topology of Z has $U \cap Z = i^{-1}(U)$ as open sets, so that the inclusion map i is continuous. It has no other opens, so it is the topology with the *fewest* open sets needed to make i continuous.

Now let X be a set and $\sim$ an equivalence relation on X. Consider the set $Y = X/\sim$ of equivalence classes of $\sim$. That is, in Y, if I write $\bar{x}$ for the class of x, I have $\bar{x} = \bar{y}$ if and only if $x \sim y$, so that Y is obtained by identifying or 'glueing together' points x and y when $x \sim y$. Every surjective map $f\colon X \to Y$ of X to a set Y is obtained in this way, by just declaring $\sim$ to be the relation $x \sim y \iff f(x) = f(y)$.

Now suppose that X is a topological space, and let $\sim$ and $f\colon X \twoheadrightarrow Y = X/\sim$ be as before. The *quotient topology* of Y has open sets defined by

$$U \subset Y \text{ is open} \iff f^{-1}(U) \text{ is open in } X.$$

It is easy to see that this satisfies the axioms for a topology. Clearly f is continuous, and this is the topology with the *most* open sets for which f is continuous. It often happens that the quotient topology of Y is not a metric topology, as we see presently.

As above, let X be a topological space, and $\sim$ an equivalence relation.

Proposition   *The quotient space $Y = X/\sim$ has the following properties.*

(1)   There is a continuous map $f \colon X \to Y$ such that

$$x \sim y \implies f(x) = f(y)$$

(that is, f is constant on equivalence classes of $\sim$).

(2)   Given a space Z and a continuous map $g \colon X \to Z$ that is constant on equivalence classes of $\sim$, there exists a unique continuous map $h \colon Y \to Z$ such that $g = h \circ f$.

Proof   (1) comes from the definition as I discussed above.

(2) Given g, the map h must take $f(x) \in Y$ to $g(x)$. In other words, an element of Y is an equivalence class $[x]$ of elements of X under $\sim$, so choose x in that class, and set $h([x]) = g(x)$. This is well defined because of the assumption that g is constant on equivalence classes. Why is h continuous? For $U \subset Z$ open, $g^{-1}(U)$ is open in X, so that $f^{-1}(h^{-1}(U))$ is open in X, and $h^{-1}(U)$ is open in Y by definition of the quotient topology of Y.   QED

This property of the topological space Y and the quotient map $f \colon X \to Y$ is called a *universal mapping property* or *UMP*. Constructions throughout abstract math can be specified in terms of UMPs: you say what you want to do (in this case, find a continuous map that is constant on equivalence classes), and then ask for the solution of a UMP. In the present case, the universal mapping property says that f does not do anything that is not forced by the conditions that f is constant on equivalence classes of $\sim$, and is continuous. In other words, f identifies exactly the equivalence classes of $\sim$, and makes no more identifications, and Y has the *most* open sets subject to f being continuous. It is interesting to analyse the above proof to see that this is exactly what is required to make h well defined and continuous.

7.6   Standard examples of glueing

The quotient topology on $X/\sim$ provides the definition of 'glueing', the space obtained from X by glueing together points $x \sim y$. Here I discuss some basic examples; see Exercises 7.18–7.19 below for more.

Example 1 $S^1 = [0, 1]/\sim$ where $\sim$ glues the endpoints (see Figure 7.2b).

Example 2   Let X be the unit square $[0, 1] \times [0, 1]$. The Möbius strip M is defined by glueing some of the sides of X as in Figure 7.6a. More formally, consider the

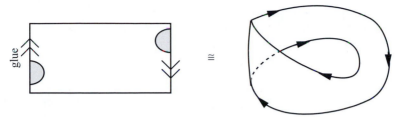

Figure 7.6a   The Möbius strip M.

Figure 7.6b   The cylinder $S^1 \times [0, 1]$.

following equivalence relations on X:

$$(x, y) \sim (x', y') \iff \begin{cases} \text{either } (x, y) = (x', y') \\ \text{or } x = 0, x' = 1 \text{ and } y = 1 - y' \\ \text{or vice versa,} \end{cases}$$

and define the Möbius strip M by $M = X/\!\!\sim$, with the quotient topology. By definition of the quotient topology, a point on the glued line has a neighbourhood obtained from neighbourhoods of its two inverse images in X.

Example 3   The cylinder $S^1 \times [0, 1]$ is obtained by glueing the unit square $[0, 1] \times [0, 1]$ as in Figure 7.6b.

Example 4   The torus $T \simeq S^1 \times S^1$ is obtained from the unit square $[0, 1] \times [0, 1]$ by the glueing of Figure 7.6c. By definition of the quotient topology, the four corners of the square correspond to a point of the torus, and a neighbourhood of it is obtained from neighbourhoods of the four corners in X. You can regard this as a surface of rotation in $\mathbb{R}^3$, or the surface in $\mathbb{R}^4$ given by $x_1^2 + y_1^2 = x_2^2 + y_2^2 = 1$.

Example 5   The surface with g handles. The picture is as in Figure 7.6d: you get it by starting from S^2, marking $2g$ distinct points on S^2, cutting out small discs around these, and glueing back in g small cylinders. See Exercise 7.19 as well as 9.4 for further discussion.

Notice that all these spaces can easily be made into metric spaces, but you do not really gain anything by doing so.

Proposition   *The Möbius strip M, the cylinder N $= S^1 \times [0, 1]$ and the torus T are not homeomorphic.*

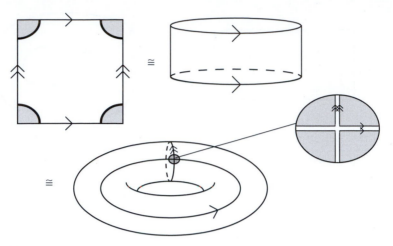

Figure 7.6c　The torus.

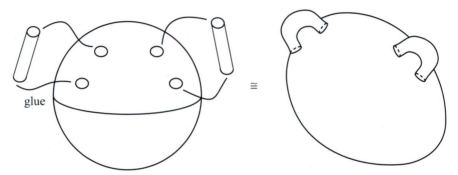

Figure 7.6d　Surface with g handles.

I can almost prove this now, though I relegate one crucial statement to the end of the chapter. The proof consists of the following steps.

Step 1.　Main claim　Points of the boundary $\partial M \subset M$ and $\partial N \subset N$ are distinguished from points of the interior by their topological properties.

Step 2　Therefore, if there exists a homeomorphism $\varphi \colon M \to N$, it must map ∂M to ∂N, and the restriction must define a homeomorphism $\partial M \simeq \partial N$.

Step 3　∂M is path connected, whereas ∂N is disconnected; hence a homeomorphism $M \simeq N$ as in Step 2 cannot exist. In the same way, $\partial T = \emptyset$, so that $T \not\simeq M$ and $T \not\simeq N$.

Given the main claim, Steps 2–3 are obvious, and the point is therefore to understand Step 1. How do I distinguish points of the interior of a surface from points on the boundary? The point is that every small neighbourhood $U \setminus P$ of an interior point P contains a small punctured disc D^* about P; the *punctured disc* is the topological

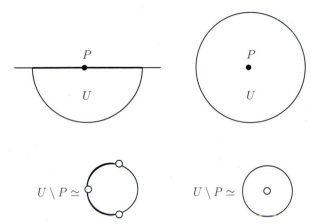

Figure 7.6e   Boundary and interior points.

space $\{0 < x^2 + y^2 < 1\} \subset \mathbb{R}^2$. On the other hand, if P is a boundary point, it has an arbitrarily small neighbourhood homeomorphic to a closed half-disc, that can be written in polar coordinates

$$U \simeq \big\{(r, \theta) \,\big|\, 0 \leq r < 1, \ \theta \in [-\pi/2, \pi/2]\big\}$$

with P at the centre of the half-disc. Hence $U \setminus P$ is homeomorphic to

$$U \setminus P \simeq \big\{(r, \theta) \,\big|\, 0 < r < 1, \ \theta \in [-\pi/2, \pi/2]\big\}$$

which in turn is homeomorphic to a closed disc with parts of the boundary removed, as in Figure 7.6e. Hence the essential content of telling interior and boundary points apart consists in showing that the punctured disc D^* is not homeomorphic to the disc D. Think it through yourself to see whether you find this statement intuitive; see 7.15.4, Corollary 1 for the proof.

7.7   Topology of $\mathbb{P}^n_\mathbb{R}$

Recall 5.2: projective n-space, as a set, is defined to be the set of lines of $\mathbb{R}^{n+1}$ through the origin, or in other words, the quotient of $\mathbb{R}^{n+1} \setminus \{0\}$ by the equivalence relation which identifies $\mathbf{x}$ with $\lambda\mathbf{x}$ for $\lambda \neq 0$. The topology of $\mathbb{P}^n$ is the quotient topology of $\mathbb{R}^{n+1} \setminus \{0\}$. This section considers various ways of looking at this topology.

Write $S^n = \{\mathbf{x} \in \mathbb{R}^{n+1} \mid \sum x_i^2 = 1\} \subset \mathbb{R}^{n+1}$ for the n-sphere. Obviously S^n meets every line of $\mathbb{R}^{n+1}$ through 0 in a pair of antipodal points. Therefore, as a set, $\mathbb{P}^n_\mathbb{R} = S^n/\pm$, where $\pm$ is the equivalence relation identifying antipodal points of the sphere (that is, pairs $\pm\mathbf{x}$ of opposite points). The topology of $\mathbb{P}^n$ coincides with the quotient topology of $S^n/\pm$; indeed, a subset of the lines through 0 is open in $\mathbb{R}^{n+1} \setminus 0$ if and only if its intersection with S^n is open in the subspace topology of S^n. Note that $S^n \subset \mathbb{R}^{n+1}$ is closed and bounded hence compact (Example 7.4.2); thus $\mathbb{P}^n$, being the continuous

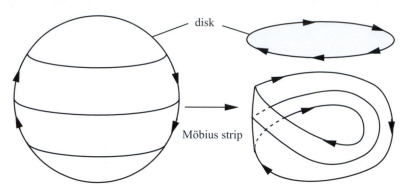

Figure 7.7   Topology of $\mathbb{P}^2_{\mathbb{R}}$: Möbius strip with a disc glued in.

image of a compact space, is also compact by the tautological Proposition 7.4.3. This was one of the motivations for constructing projective space discussed in 5.1.4.

There are many ways of understanding the quotient, by choosing a closed subset of S^n that picks out just one of each pair of antipodal points for a big open subset and then glueing around the boundary: for example, the closed northern hemisphere of S^n contains one of each pair of antipodal points, except that I still have to identify antipodal points of the equatorial sphere S^{n-1}.

In the case $n = 2$, we can do the following: view S^2 as the union of 3 pieces, a cap around the north pole, a band around the equator, and a cap around the south pole (see Figure 7.7). Every point in the southern cap is equivalent to a point in the northern cap, so the southern cap is not needed. Now cut the equatorial band into its front and back halves; as before, every point in the back half is equivalent to a point in the front half, so this piece is also not needed. Now $\pm$ glues together the left and right intervals of the front half to give a Möbius strip; this glueing is the same as in Figure 7.6a. The northern cap is a disc, with boundary a circle; the Möbius strip also has boundary a circle, and $\mathbb{P}^2$ is obtained by glueing these two pieces together along their boundaries. Note that this is an abstract construction: you cannot do it in $\mathbb{R}^3$ without allowing self-crossing.

It is an interesting exercise to see the components of this construction as the result of cutting $\mathbb{P}^2$ along a line and along a conic. See Exercise 7.17(a).

7.8   Nonmetric quotient topologies

Example 1 (The mousetrap topology) $X = \{P, Q\}$ is a space with only 2 points and open sets

$$\mathcal{T}_X = \Big\{ \emptyset, \{P\}, X \Big\}.$$

Here P is an open point, but not Q. Every neighbourhood of Q (there is only one) contains P. In terms of convergence, the constant sequence $P, P, \ldots$ converges both to P and to Q (please check this as an instant exercise; refer back to 7.4.2 for the

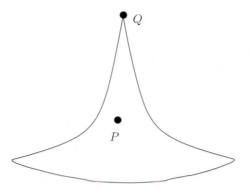

Figure 7.8a   The mousetrap topology.

definition of convergence if needed). This implies, of course, that the topology of X is not metric. X is a quotient topology: introduce the equivalence relation $\sim$ on $\mathbb{C}$ defined by

$$x \sim y \iff x = \lambda y \quad \text{with } \lambda \in \mathbb{C}, \lambda \neq 0.$$

Then there are only two equivalence classes, $Q = [0]$ and $P = [\lambda \in \mathbb{C} \setminus \{0\}]$; $\{P\}$ is obviously open while $\{Q\}$ is not. The point is that if you are at 0 then any arbitrarily small perturbation takes you into a nonzero number; that is, viewed from Q, the point P is infinitely close. But if you are at a nonzero number λ, all the points in a small neighbourhood are also nonzero, so viewed from P, the point Q is far away. Being zero is an unstable, or closed condition; being nonzero is a stable or open condition.

I call this the *mousetrap topology* (Figure 7.8a) because if you are at Q (outside the trap), it is no distance at all to get into the trap. But if you are at P (inside the trap), then it is a long way out. Thus the content of the topology is more logical than geometric.

There are many equivalence relations of interest with this kind of behaviour. One example is the equivalence relation on $\mathbb{R}$ with

$$\{x \in \mathbb{R} \mid x > 0\}, \quad \{0\}, \quad \{x \in \mathbb{R} \mid x < 0\}$$

as its 3 equivalence classes.

Example 2 (Quadratic forms)   A similar but more substantial example: consider quadratic forms $q(x, y) = ax^2 + 2bxy + cy^2$ on $\mathbb{R}^2$. There is a coordinate change that puts $q(x, y)$ in one of the 6 normal forms:

$$q_1 = x^2 + y^2, \ \ q_2 = x^2 - y^2, \ \ q_3 = -x^2 - y^2, \ \ q_4 = x^2, \ \ q_5 = -x^2, \ \ \text{or } q_0 = 0.$$

All the quadratic forms on $\mathbb{R}^2$ are parametrised by $(a, b, c) \in \mathbb{R}^3$, corresponding to the symmetric matrix $A = \left(\begin{smallmatrix} a & b \\ b & c \end{smallmatrix} \right)$. Now introduce the equivalence relation on $\mathbb{R}^3$

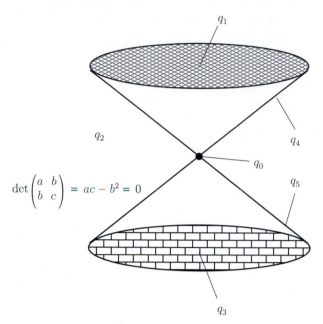

$$\det \begin{pmatrix} a & b \\ b & c \end{pmatrix} = ac - b^2 = 0$$

Figure 7.8b   Equivalence classes of quadratic forms $ax^2 + 2bxy + cy^2$.

corresponding to a coordinate change:

$$A \sim B \iff \exists M \in \mathrm{GL}(2, \mathbb{R}) \text{ such that } A = {}^t MBM.$$

(Here $\mathrm{GL}(2, \mathbb{R})$ is the group of 2×2 invertible matrixes.) This means exactly that I consider quadratic forms up to change of basis. So there are exactly the 6 classes, the strata of Figure 7.8b. The quotient topology on the set

$$X = \mathbb{R}^3/\!\sim = \{q_1, q_2, q_3, q_4, q_5, q_0\}$$

has open sets

$$\{q_1\}, \quad \{q_2\}, \quad \{q_3\}, \quad \{q_4, q_1, q_2\}, \quad \{q_5, q_2, q_3\}, \quad X$$

and their unions. For example, every neighbourhood of q_4 contains q_1, q_2.

7.9   Basis for a topology

This is a formal idea for constructing topologies. Let $\mathcal{B}$ be a collection of subsets of X. Then $\mathcal{B}$ is a *basis for a topology* if it satisfies the three axioms

1. finite intersections:

$$U_1, \ldots, U_n \in \mathcal{B} \implies U_1 \cap \cdots \cap U_n \in \mathcal{B};$$

2. involves every point: for all $x \in X$ there exists $U \in \mathcal{B}$ such that $x \in U$;
3. empty convention: $\emptyset \in \mathcal{B}$.

Construction   If $\mathcal{B}$ is a basis for a topology, the family of subsets

$$\mathcal{T} = \left\{ \bigcup_{\lambda \in \Lambda} U_\lambda : U_\lambda \in \mathcal{B}, \Lambda \text{ arbitrary index set} \right\}$$

of X is a topology on X, the topology generated by $\mathcal{B}$.

Proof   This is entirely formal. $X \in \mathcal{T}$ using axiom 2 and the construction. $\mathcal{T}$ is closed under arbitrary unions by construction. To show that $\mathcal{T}$ is closed under finite intersections, note that

$$\left(\bigcup_{\lambda \in \Lambda} U_\lambda \right) \cap \left(\bigcup_{\mu \in M} U_\mu \right) = \bigcup_{\lambda, \mu} U_\lambda \cap U_\mu. \quad \text{QED}$$

I can save time by listing only a basis for the topology, rather than by saying what all the open sets are. The idea here is that a topology is specified by the neighbourhoods of each point (because an open set is determined by the condition that it is a neighbourhood of each of its points). In turn it is enough to specify any system of sufficiently small neighbourhoods of each point.

Example 1   In 7.8, Example 2, I described the quotient topology on $X = \mathbb{R}^3/\sim$ by telling you that its open sets are unions of

$$\{q_1\}, \quad \{q_2\}, \quad \{q_3\}, \quad \{q_4, q_1, q_2\}, \quad \{q_5, q_2, q_3\}, \quad \{q_0, q_1, q_2, q_3, q_4, q_5\}.$$

Example 2   Let X, d be a metric space, and

$$\mathcal{B} = \left\{ B(x_1, \varepsilon_1) \cap \cdots \cap B(x_n, \varepsilon_n) \right\}$$

be the set of finite intersection of open balls $B(x, \varepsilon) = \{y \mid d(x, y) < \varepsilon\}$. Then $\mathcal{B}$ is a basis for a topology $\mathcal{T}$, the usual metric topology.

Example 3.   Profinite topology of an infinite group   Another more substantial example. Take any group G; recall that a subgroup $H \subset G$ is normal (written $H \lhd G$) if $gH = Hg$ for every $g \in G$, that is, its right and left cosets coincide. A normal subgroup $H \lhd G$ of finite index n is the kernel of a surjective homomorphism $G \to \Gamma$ to a finite group of order n. For example, if $G = \mathbb{Z}$ then every normal subgroup of finite index is just $n\mathbb{Z}$ for some integer n.

Let G be a group, with $e \in G$ the identity element. Then there is a topology on G such that:

(a)   normal subgroups $H \lhd G$ of finite index form a set of sufficiently small neighbourhoods of e;

(b)   the right translation maps $r_g \colon G \to G$ defined by $f \mapsto fg$ are homeomorphisms.

It follows from (a) and (b) that a set of sufficiently small neighbourhoods of any $g \in G$ are given by cosets gH, where the H are as in (a). So take

$$\mathcal{B} = \{\emptyset\} \cup \{\text{cosets of normal subgroups of finite index}\}$$

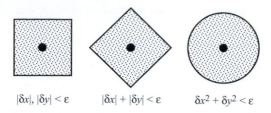

$$|\delta x|, |\delta y| < \varepsilon \qquad |\delta x| + |\delta y| < \varepsilon \qquad \delta x^2 + \delta y^2 < \varepsilon$$

Figure 7.10   Balls for product metrics.

as a basis for a topology. I check that this is a basis by going through the three axioms. Indeed, $\emptyset, G \in \mathcal{B}$. Also if $H_1, \ldots, H_n$ are normal subgroups of finite index then so is $H_1 \cap \cdots \cap H_n$, clearly, and if $g_1 H_1, \ldots, g_n H_n$ are their cosets then either $g_1 H_1 \cap \cdots \cap g_n H_n = \emptyset$, or $\exists g \in g_1 H_1 \cap \cdots \cap g_n H_n$, in which case

$$g_1 H_1 \cap \cdots \cap g_n H_n = g H_1 \cap \cdots \cap g H_n = g(H_1 \cap \cdots \cap H_n).$$

The topology generated by this basis is called the *profinite topology* of G. Note that if $H \lhd G$ is a normal subgroup of finite index then its cosets form a partition of G by finitely many disjoint open sets. Therefore any of these cosets is also closed.

Remark   Profinite topologies on groups have lots of applications in algebra and number theory. For example, in number theory, you may want to solve an equation $f(x, y) = 0$ in $\mathbb{Z}$, knowing that you can solve it modulo all N. Another example occurs in Galois theory. The idea is that if $k \subset L$ is an infinite Galois field extension, the finite extension fields $k \subset K \subset L$ correspond to subgroups of finite index in the infinite Galois group $\mathrm{Gal}(L/k)$. The Galois group $\mathrm{Gal}(L/k)$ is automatically profinite, in the sense that it is defined by its finite quotient groups.

7.10   Product topology

Let X and Y be topological spaces; I show how to put a topology on $X \times Y$. Take the set of subsets

$$\mathcal{B} = \{U \times V \subset X \times Y\} \quad \text{with } U \subset X \text{ and } V \subset Y \text{ open.}$$

Then

$$(U_1 \times V_1) \cap (U_2 \times V_2) = (U_1 \cap U_2) \times (V_1 \cap V_2)$$

gives the finite intersection property; the other two axioms are obvious, so $\mathcal{B}$ is a basis for a topology on $X \times Y$. The *product topology* on $X \times Y$ is defined to be the topology generated by $\mathcal{B}$.

If X and Y are metric spaces, it is easy to see that the product topology on $X \times Y$ is the topology defined by any of the metrics $\max(d_X, d_Y)$, $d_X + d_Y$, $\sqrt{d_X^2 + d_Y^2}$, etc. (see Figure 7.10). It follows that for n, m positive integers, the product topology on $\mathbb{R}^n \times \mathbb{R}^m$ is the same as the metric topology on $\mathbb{R}^{n+m}$. For example, on $\mathbb{R}^2 = \mathbb{R} \times \mathbb{R}$, the sets $(a_1, b_1) \times (a_2, b_2)$ provide arbitrarily small open sets, but obviously not all open sets are of this form.

7.11 The Hausdorff property

A topological space is *Hausdorff*[1] if for all $x \neq y \in X$, there exist disjoint open sets $U, V \subset X$ with $x \in U$, $y \in V$. (See Figure 7.2a.) This is clearly another topological property.

If X is Hausdorff then every point $x \in X$ is closed: for if $x \neq y$ there exists an open set containing y and not x, and therefore $X \setminus x$ is open. This is a weaker separation axiom,

$$\forall x \neq y \in X, \ \exists \text{ an open set } U \text{ containing } y \text{ and not } x$$

called Hausdorff's T_1 condition. (The Hausdorff condition on X introduced here is sometimes also called T_2.)

Example 1   A metric space X is automatically Hausdorff: just choose $\varepsilon > 0$ with $\varepsilon < \frac{1}{2} d(x, y)$ and set $U = B(x, \varepsilon)$, $V = B(y, \varepsilon)$.

Example 2   Examples 1 and 2 of 7.8 are clearly not Hausdorff. The cofinite topology of an infinite set X (7.1 Example 2) is not Hausdorff either: a nonempty open set is the complement of a finite set, so the intersection of any two open sets is again the complement of a finite set, so nonempty. Thus these are certainly not metric topologies.

Example 3   A topology on a finite set X is Hausdorff if and only if it is the discrete topology. Indeed, if X is Hausdorff then any point $x \in X$ is closed, so every subset of X is closed.

Proposition   *A topological space X is Hausdorff if and only if the diagonal*

$$\Delta_X = \{(x, x) \mid x \in X\} \subset X \times X$$

is closed in the product topology of X.

Proof   Note first that for any subsets $U, V \subset X$,

$$U \times V \cap \Delta_X = \{(x, x) \mid x \in U \cap V\},$$

in other words, $U \times V \cap \Delta_X$ is just the diagonal embedding of $U \cap V$ into $X \times X$.

A point of $X \times X \setminus \Delta_X$ is just a pair (x, y) with $x \neq y$. Consider the problem of finding an open neighbourhood W of (x, y) in the product topology such that

---

[1]   Felix Hausdorff (1868–1942) was the originator of many of the basic ideas of metric and topological spaces, and the author of a famous and influential book *Grundzüge der Mengenlehre*. He was Professor at the University of Bonn until he was forced out as a Jew in 1935. He committed suicide in January 1942, together with several members of his family, to avoid being sent to a Nazi internment camp.

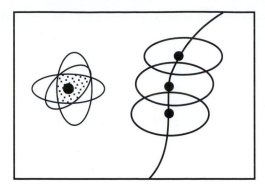

Figure 7.12   Separating a point from a compact subset.

$W \cap \Delta_X = \emptyset$. By definition of the product topology, an arbitrary small neighbourhood of (x, y) is $U \times V$ with $U, V \subset X$ open and $x \in U$, $y \in V$. Now by the first remark, $U \times V \cap \Delta_X = \emptyset$ if and only if $U \cap V = \emptyset$.

Since Δ_X is closed if and only if $X \times X \setminus \Delta_X$ is open, this happens if and only if for every (x, y) with $x \neq y$ there exist open sets $U, V \subset X$ open, with $x \in U$, $y \in V$ and $U \cap V = \emptyset$.   QED

7.12   Compact versus closed

Proposition   *Let X be a topological space, and $Y \subset X$ a subset with the subspace topology.*

(i)   *If X is a compact topological space and $Y \subset X$ is closed, then Y is also compact.*
(ii) *If X is Hausdorff and $Y \subset X$ is compact, then Y is closed.*
(iii) *In particular, if X is compact and Hausdorff, then $Y \subset X$ is compact if and only if it is closed.*

Proof   (i) Suppose that V_λ for $\lambda \in \Lambda$ are open subsets of Y, in the subspace topology, such that $Y = \bigcup V_\lambda$. Then by definition of the subspace topology 7.5, for each λ there exists an open set U_λ of X such that $V_\lambda = Y \cap U_\lambda$. Now also $X \setminus Y$ is open, by the assumption that Y is closed. Therefore $X = \bigcup U_\lambda \cup (X \setminus Y)$ is an open cover of X. By definition of compactness, a finite cover will do, say $X = \bigcup_{i=1}^n U_{\lambda_i} \cup (X \setminus Y)$, and then obviously $Y = \bigcup_{i=1}^n V_{\lambda_i}$.

(ii) Fix $x \in X \setminus Y$. For every $y \in Y$, using the Hausdorff assumption on X, choose disjoint open sets U_y and V_y with $x \in U_y$ and $y \in V_y$. By construction, $y \in V_y$, so that $Y \subset \bigcup V_y$, or equivalently $Y = \bigcup (Y \cap V_y)$. But since Y is compact, a finite number of the open sets $Y \cap V_y$ cover it, and hence there is a finite set of V_{y_i} with $Y \subset \bigcup_{i=1}^n V_{y_i}$. Set $U = \bigcap_{i=1}^n U_{y_i}$, which is a finite intersection of opens, therefore open. Since $U_y \cap V_y = \emptyset$ for each y, it follows that $U \cap \bigcup_{i=1}^n V_{y_i} = \emptyset$, and in particular $U \cap Y = \emptyset$. (See Figure 7.12.) This proves that for any $x \notin Y$, there exists an open set U containing x disjoint from Y, and therefore Y is closed.   QED

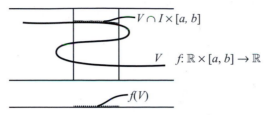

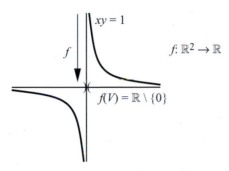

Figure 7.13a   Closed map.

Figure 7.13b   Nonclosed map.

7.13   Closed maps

A map $f : X \to Y$ between topological spaces is *closed*, if $f(V) \subset Y$ is closed for every closed set $V \subset X$.

Example 1   Consider the closed interval $[a, b] \subset \mathbb{R}$. Then the second projection $\pi : [a, b] \times \mathbb{R} \to \mathbb{R}$ is a closed map (Figure 7.13a).

Proof   Start with a closed set $V \subset [a, b] \times \mathbb{R}$ and a point $x \in \overline{\pi(V)}$ of the closure of $\pi(V)$. Take a closed interval I containing x, and restrict attention to the second projection

$$B = [a, b] \times I \to I.$$

Then B is closed and bounded in $\mathbb{R}^2$, so compact (see Proposition 7.4.2); hence $V \cap B$ is compact by Proposition 7.12 (i). Therefore by Proposition 7.4.3, $f(V \cap B)$ is a compact subset of I, therefore closed in I. Therefore $x \in \pi(V)$, and $\pi(V)$ is closed.   QED

Example 2   The projection to the x-axis $\mathbb{R}^2 \to \mathbb{R}$ is not closed. For consider the hyperbola $C : (xy = 1)$; it is closed in $\mathbb{R}^2$, but its image in $\mathbb{R}$ is $\mathbb{R} \setminus 0$ (Figure 7.13b).

Proposition   *If X is compact and Y Hausdorff then any continuous map $f : X \to Y$ is closed.*

Proof $V \subset X$ closed implies V compact by Proposition 7.12 (i). Therefore $f(V)$ is compact by Proposition 7.4.3, and $f(V) \subset Y$ is closed by Proposition 7.12 (ii).   QED

7.14   A criterion for homeomorphism

Let X and Y be topological spaces and $f : X \to Y$ a map. I claim that

$$f \text{ is a homeomorphism } \iff f \text{ is bijective, continuous, and closed.}$$

$\implies$ is of course clear. If f is bijective, then f closed means exactly that f^{-1} is continuous: for $U \subset X$ open gives $X \setminus U$ closed, which implies that $f(X \setminus U)$ is closed; but $f(X \setminus U) = Y \setminus f(U)$ because f is bijective, so $f(U)$ is open, that is, f^{-1} is continuous.

Theorem ($\heartsuit = 0$)   *If X is compact and Y Hausdorff, then a continuous bijective map $f : X \to Y$ is a homeomorphism.*

Proof f is closed by Proposition 7.13.   QED

Example   A *simple closed curve* in $\mathbb{R}^2$ is a continuous map $f : [0, 1] \to \mathbb{R}^2$ that is one-to-one except for $f(0) = f(1)$. Write $\sim$ for the equivalence relation that glues the endpoints of the interval as in Figure 7.2b. Clearly f defines a continuous one-to-one map $f' : [0, 1]/\sim = S^1 \to \mathbb{R}^2$. I claim that $f' : S^1 \to f'(S^1)$ is a homeomorphism. Indeed, it is a continuous one-to-one map from a compact space S^1 to a Hausdorff space $f'(S^1) \subset \mathbb{R}^2$. This proves that $\heartsuit = 0$.

7.15   Loops and the winding number

Let $D = \{(x, y) \in \mathbb{R}^2 \mid x^2 + y^2 < 1\}$ be the unit disc in $\mathbb{R}^2$ and $D^* = D \setminus (0, 0)$ the punctured disc. This final section will answer the following question, left open in the proof of Proposition 7.6.

Question   How can we tell that D^* is not homeomorphic to D?

Answer D is *simply connected*: any loop in D (starting and ending at P_0, say) can be contracted in D to the constant loop; on the other hand, a loop in D^* has a *winding number n* around the puncture $(0, 0)$, and the loop can be contracted if and only if $n = 0$.

The intuitive picture is clear: think of taking a dog on a long lead for a walk in a park having a tall pole in the middle. In classical math, the winding number n is the ambiguity of $2\pi n$ in the functions $\arcsin x$ and $\arccos x$ and the ambiguity of $n(2\pi i)$ in the complex function $\log z$. The content of the following sections is the first step in the theory of the fundamental group $\pi_1(X, P_0)$ in algebraic topology; Theorem 7.15.3 on the winding number is closely related to the statement that $\pi_1(D^*, P_0) = \mathbb{Z}$.

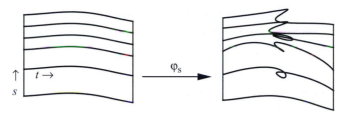

Figure 7.15a   Continuous family of paths.

7.15.1
Paths, loops
and families

Recall that a path in a topological space X is a continuous map $\varphi \colon [0, 1] \to X$, written $t \mapsto \varphi(t)$. Fix a *base point* $P_0 \in X$. A *loop in X based at P_0* is a path starting and ending at P_0; in other words, a continuous map $f \colon [0, 1] \to X$ such that $f(0) = f(1) = P_0$. These are called *based loops* (as opposed to *free loops* where we insist that $f(0) = f(1)$, but allow this to be any point in X). A loop is allowed to cross over itself any number of times, or even to stop for a while or go back along itself.

A *family of paths (or loops)* $(\varphi^{(s)})$ depending on a parameter $s \in [0, 1]$ is just an indexed family of paths (or loops), one for each $s \in [0, 1]$. Write I_t for the interval $[0, 1]$ of the path parameter t, and I_s for the interval $[0, 1]$ of the family parameter s.

Tentative definition   Let X be a metric space. A family of paths $(\varphi^{(s)})$ is *continuous* at s if for every $\varepsilon > 0$, there exists a δ such that

$$|s - s'| < \delta \implies d(\varphi^{(s)}(t), \varphi^{(s')}(t)) < \varepsilon \quad \text{for all } t \in [0, 1].$$

We say that $(\varphi^{(s)})$ is a *continuous family of paths* if it is continuous at all $s \in [0, 1]$. The definition applies in exactly the same way to a family of based loops, except that I insist that $\varphi^{(s)}(0) = \varphi^{(s)}(1) = P_0$ for every s.

Note that the continuity assumption is uniform in t (the same δ is supposed to guarantee closeness for all t). The hard thing is to understand why the definition just given is the right one. The point is that to say that the path $\varphi^{(s)}$ moves just a little, we have to guarantee that every step $\varphi^{(s)}(t)$ for fixed t should move just a little, bounded in t (compare Exercise 7.20).

Lemma   *Corresponding to a family of paths $(\varphi^{(s)})$, consider the map*

$$\Phi \colon I_s \times I_t = [0, 1] \times [0, 1] \to X \quad \text{given by} \quad \Phi(s, t) = \varphi^{(s)}(t).$$

Then $(\varphi^{(s)})$ is a continuous family of paths if and only if Φ is continuous. See Figure 7.15a.

Remark   Notice that Φ continuous is a topological property. The point of the lemma is that it makes the notion of continuous family of paths purely topological. If X is a topological space, the 'uniform' definition of a continuous family of paths is not applicable (it depends on the metric in X); in the Definition below I define a family of paths $\varphi^{(s)}$ to be continuous by the property that Φ is continuous.

Proof $\Longrightarrow$ A standard 'divide the ε in two' argument. Suppose we are given $(s_0, t_0) \in I_s \times I_t$ and $\varepsilon > 0$. First, because $\varphi^{(s_0)}$ is continuous, there exists δ such that

$$d(t, t_0) < \delta \implies d(\varphi^{(s_0)}(t), \varphi^{(s_0)}(t_0)) < \varepsilon/2.$$

Next, because $\varphi^{(s)}$ is a continuous family of paths at s_0, there exists a δ such that

$$d(s, s_0) < \delta \implies d(\varphi^{(s)}(t), \varphi^{(s_0)}(t)) < \varepsilon/2 \quad \text{for all } t.$$

Therefore $\max\{d(s, s_0), d(t, t_0)\} < \delta$ implies both of these inequalities, so that

$$d((s, t), (s_0, t_0)) = \max\{d(s, s_0), d(t, t_0)\} < \delta \implies$$
$$d(\Phi(s, t), \Phi(s_0, t_0)) \le d(\varphi^{(s_0)}(t), \varphi^{(s_0)}(t_0)) + d(\varphi^{(s)}(t), \varphi^{(s_0)}(t)) < \varepsilon.$$

This proves Φ is continuous as a function of (s, t).

$\Longleftarrow$ In this direction, I have to use compactness of I_t to get uniformity in t. If Φ is continuous, each $\varphi^{(s)} \colon I_t \to X$ is obviously continuous. I fix some $s_0 \in I_s$, and try to prove that $(\varphi^{(s)})$ is a continuous family of paths at s_0. Suppose given $\varepsilon > 0$. Start by working in a neighbourhood of a fixed $t \in I_t$.

Then because Φ is continuous at (s_0, t), there exists some δ (possibly depending on t) such that

$$d((s, t'), (s_0, t)) < \delta \implies d(\varphi^{(s)}(t'), \varphi^{(s_0)}(t)) < \varepsilon/2.$$

Therefore $d(s, s_0) < \delta$ and $d(t', t) < \delta$ implies that $\varphi^{(s)}(t')$ is close to $\varphi^{(s_0)}(t)$ is close to $\varphi^{(s_0)}(t')$. In other words, for all t', there is a δ neighbourhood of t,

$$d(s, s_0) < \delta \implies d(\varphi^{(s)}(t'), \varphi^{(s_0)}(t')) < \varepsilon.$$

Now I have proved that every point of the t-interval has a δ neighbourhood with this property; by compactness the t-interval is covered by finitely many of these, and by taking δ to be the minimum of finitely many δ_i I get $\varphi^{(s)}(t)$ close to $\varphi^{(s_0)}(t)$ for all t and all s close to s_0.   QED

Definition   Let X be a topological space and $P_0 \in X$ a base point. A family of loops $\varphi^{(s)}$ in X based at P_0 is *continuous*, if the map

$$\Phi \colon [0, 1] \times [0, 1] \quad \text{defined by} \quad \Phi(s, t) = \varphi^{(s)}(t)$$

is continuous. A loop $\varphi \colon [0, 1] \to X$ based at P_0 is *contractible* in X, if there is a continuous family of loops joining φ to the constant loop φ_0 (defined by $\varphi_0(t) = P_0$ for all t). A path connected space X is *simply connected* if every loop in X (with every possible base point, though see Exercise 7.21) is contractible.

A homeomorphism $f \colon X \to Y$ takes paths and continuous families of paths in X into paths and continuous families of paths in Y. In particular, being simply connected is a topological property.

Example   Every loop in the unit disc $D \in \mathbb{R}^2$ is contractible. This is obvious on a sheet of paper; formally, it is best to use vector notation: if $\mathbf{x}_0$ is the base point, and $\varphi(t) = \mathbf{x}_t$ is the loop then $\Phi(s, t) = \mathbf{x}_0 + s(\mathbf{x}_t - \mathbf{x}_0)$ gives a continuous family of paths connecting φ to the constant path at $\mathbf{x}_0$. The point is just that D is convex; the same argument gives the same conclusion for any convex subset of $\mathbb{R}^n$.

7.15.2
The winding
number

To discuss the winding number formally, I use ordinary Cartesian coordinates (x, y) on the disc D, and polar coordinates (r, θ) on the punctured disc D^*. Note that $r > 0$, and that polar coordinates do not really work at the origin. The two coordinate systems are related by the usual rules $x = r \sin \theta$, $y = r \cos \theta$.

What values do we allow for θ? Since sin and cos are periodic with period 2π, the right answer is an equivalence class of $\mathbb{R}$ modulo $2\pi \mathbb{Z}$. Note that every equivalence class of $\mathbb{R}/2\pi \mathbb{Z}$ has a unique representative $\theta \in [0, 2\pi)$; in applications $\theta \in (-\pi, \pi]$ may be more convenient. If you want θ to be unique, you should insist that $(x, y) \neq (0, 0)$, and choose the representative $\theta \in [0, 2\pi)$. But if you want θ to vary continuously with (x, y), you should arrange that (x, y) stays well away from $(0, 0)$ and choose $\theta \in \mathbb{R}$.

Proposition   *Suppose that the base point P_0 is in the x-axis (so that $\theta = 0$ is a possible choice). Let $\varphi \colon [0, 1] \to D^*$ be a path with $\varphi(0) = P_0$. Then there exist unique continuous functions $r \colon [0, 1] \to \mathbb{R}_+$ and $\Theta \colon [0, 1] \to \mathbb{R}$ such that*

$$\varphi(t) = (r(t), \Theta(t)) \quad \text{for all } t \in [0, 1].$$

If φ is a loop, then the end point is $\varphi(1) = P_0$; hence the value $\Theta(1)$ is of the form $2\pi n$ for some integer n.

Definition   The integer number n in the expression $\Theta(1) = 2\pi n$ is the *winding number* of the loop φ, written $n = \nu(\varphi)$.

Proof   Write $\varphi(t) = (x(t), y(t))$ and set $r(t) = \sqrt{x(t)^2 + y(t)^2}$ for $t \in [0, 1]$. Clearly $r(t)$ is continuous and strictly positive. Since $[0, 1]$ is compact, $r(t)$ is bounded above and below by some $R, \rho > 0$. Define

$$\varphi_1 \colon [0, 1] \to S^1 \quad \text{by} \quad \varphi_1(t) = \left(\frac{x(t)}{r(t)}, \frac{y(t)}{r(t)} \right).$$

Then φ_1 is continuous, because x, y and r are, and $r(t)$ is bounded away from 0. Now $\varphi_1(t) \in S^1$ is certainly of the form $(\sin \theta, \cos \theta)$ for some $\theta = \theta(t) \in \mathbb{R}$. The problem is that $\theta(t)$ is determined up to addition of multiples of 2π, and we have to choose the value for each t to make the function continuous.

Clearly the map $e \colon \mathbb{R} \to S^1$ defined by

$$e \colon \theta \mapsto (\sin \theta, \cos \theta)$$

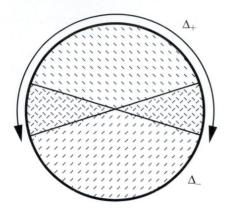

Figure 7.15b D^* covered by overlapping open radial sectors.

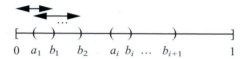

Figure 7.15c Overlapping intervals.

defines a homeomorphism of any open interval $(a, b) \subset \mathbb{R}$ of length $b - a < 2\pi$ onto an open sector of the circle S^1 (similarly for closed). To prove the proposition, it is enough to chop up $[0, 1]$ into finitely many short intervals U_i so that φ_1 maps each U_i into such a sector, then take a suitable branch of e^{-1} on each of these.

To do this very explicitly, cover D^* by a number of overlapping open radial sectors. To be definite, say, the 'top' and 'bottom' 200° sectors

$$\Delta_+ : -10° < \theta < 190°, \quad \Delta_- : 170° < \theta < 370°,$$

as in Figure 7.15b (or make your own choice). Let me write $\varepsilon = 10° = \pi/18$, so that the sector intervals are $(0 - \varepsilon, \pi + \varepsilon)$ and $(\pi - \varepsilon, 2\pi + \varepsilon)$. Then $\mathbb{R}$ is divided up into countably many intervals

$$I_+^l = (2l\pi - \varepsilon, (2l + 1)\pi + \varepsilon) \quad \text{and} \quad I_-^l = ((2l - 1)\pi - \varepsilon, 2l\pi + \varepsilon)$$

for $l \in \mathbb{Z}$, in such a way that the restriction of e to each interval $I_\pm^l$ is a homeomorphism $e_\pm^l : I_\pm^l \to \Delta_\pm$.

For every $t \in [0, 1]$, the image $\varphi_1(t) \in D^*$ is in one of the $\Delta_\pm$. Since φ_1 is continuous, $\varphi_1^{-1}(\Delta_\pm)$ is open, so there exists a neighbourhood $U(t) \subset [0, 1]$ of t with $\varphi_1(U(t)) \subset \Delta_\pm$. I can assume that each of the $U(t)$ is an open interval of $[0, 1]$ (except the first and last, which are half-open intervals). The $U(t)$ form an open cover of $[0, 1]$, so by compactness it has a finite subcover. It follows that I can choose a cover of $[0, 1]$ by a finite number of overlapping open intervals (Figure 7.15c)

$$[0, 1] = \bigcup_{i=0}^{m} U_i, \qquad \begin{array}{l} \text{with } U_0 = [0, b_1), \ U_i = (a_i, b_{i+1}), \ U_n = (a_m, 1], \text{ and} \\ 0 < a_1 < b_1 < a_2 < \cdots < b_{m-1} < a_n < b_m < 1, \end{array}$$

such that $\varphi_1(U_i) \subset \Delta_{\pm}$. (For each U_i, if there is any doubt, make the choice of $\pm$ at the outset.)

Now since $e_{\pm}^l : I_{\pm}^l \to \Delta_{\pm}$ is a homeomorphism, we clearly define Θ over $U_i \subset \Delta_{\pm}$ to be $(e_{\pm}^l)^{-1} \circ \varphi_1$, and the only remaining question is the choice of l. First, $\varphi(0) = P_0$ has $\theta = 0$ by assumption, so that either $U_0 \subset \Delta_+$ or $U_0 \subset \Delta_-$. In the first case, choose I_+^0, in the second choose I_-^0. These are forced by the requirement that $\Theta(0) = 0$. Next, suppose by induction that Θ is defined and continuous on $U_0 \cup U_1 \cup \cdots \cup U_{i-1}$. The initial point a_i of U_i is in the overlap with U_{i-1}, so that Θ is already defined there. This determines the choice of $I_{\pm}^l$.   QED

7.15.3 **Winding** **number is** **constant in** **a family**	**Theorem** *Let $(\varphi^{(s)})$ be a continuous family of loops $\varphi^{(s)} : [0, 1] \to D^*$. Then the winding number of the loop $\varphi^{(s)}$ is constant (independent of s). In particular $\nu(\varphi^{(0)}) = \nu(\varphi^{(1)})$.*

Proof   Write $\nu(\varphi)$ for the winding number of a loop φ. The point is to show that $\nu(\varphi)$ depends continuously on the path $\varphi : [0, 1] \to D^*$.

For some value s, suppose that $\nu(\varphi^{(s)}) = n$. I claim that there is a neighbourhood $V_s = (s - \delta, s + \delta)$ such that $\nu(\varphi^{(s')}) = n$ for all $s' \in V_s$. In other words, the subset

$$\Sigma_n = \left\{ s \mid \nu(\varphi^{(s)}) = n \right\} \subset [0, 1]$$

is open. This claim proves the theorem, because the interval $[0, 1]$ is connected, and is a disjoint union of the open sets Σ_n, therefore only one value of n occurs.

First, as in the proof of Proposition 7.15.2, I normalise all the paths by dividing by the factor $r^{(s)}(t)$, so that each $\varphi^{(s)}$ maps to S^1. The normalisation factor is bounded away from 0 because $I_s \times I_t = [0, 1] \times [0, 1]$ is compact and $\Phi : I_s \times I_t \to D^*$ is continuous. Thus I assume from now on that $\varphi^{(s)} : [0, 1] \to S^1$.

Recall the construction of Proposition 7.15.2 for $\varphi^{(s)}$. There is a cover of $[0, 1] = I_t$ by a finite chain of overlapping open intervals $U_i = (a_i, b_{i+1})$ such that $\varphi_1^{(s)}(U_i) \subset \Delta_{\pm}$. After this, the map Θ just lifts $\Delta_{\pm}$ to $I_{\pm}^n$, where the value of n is determined inductively by the already known value of the starting point $\Theta(a_i)$.

Now I choose slightly bigger 'top' and 'bottom' sectors $\Delta_{\pm}'$ of S^1; to be explicit, choose

$$\Delta_+' : -20° < \theta < 200°, \qquad \Delta_-' : 160° < \theta < 380°,$$

or in the previous notation $\Delta_+' = (0 - 2\varepsilon, \pi + 2\varepsilon)$, etc. As far as $\varphi^{(s)}$ is concerned, nothing has changed: I still have $\varphi_1^{(s)}(U_i) \subset \Delta_{\pm} \subset \Delta_{\pm}'$, and the construction of Θ can be made equally well with the bigger intervals.

However, by the definition of continuous family of loops, there exists a small neighbourhood $s \in V_s \subset [0, 1]$ such that also $\varphi^{(s')}(U_i) \subset \Delta'_\pm$ for all $s' \in V_s$. Thus I can use the same collection of intervals U_i to construct the argument function $\Theta^{(s')}$ of $\varphi^{(s')}$ for all $s' \in V_s$.

Then $\Theta^{(s')}(t)$ on $V_s \times U_i$ is equal to the composite $e^n_\pm \circ \Phi(s', t)$, and hence it is a continuous function of $(s', t) \in V_s \times U_i$. It follows that $\Theta^{(s')}(t)$ is a continuous function of $s' \in V_s$ for any t. In particular, $\Theta^{(s')}(1)$ is a continuous function of $s' \in V_s$. However, it is an integer multiple of 2π. Therefore it is constant for $s' \in V_s$. This proves the claim.   QED

**7.15.4
Applications
of the
winding
number**

Corollary 1   *The punctured disc D^* is not homeomorphic to the disc D.*

Proof   By Theorem 7.15.3, a loop φ in D^* of winding number $\neq 0$ is not contractible. On the other hand, every loop in D is contractible (Example 7.15.1). The property that a loop is contractible is a topological property, so is preserved by homeomorphism. Therefore there does not exist a homeomorphism between D and D^*.   QED

Remark   The same proof shows that the punctured disc D^* is not homeomorphic to the disc D with some of its boundary added, since loops in the latter are still contractible. This concludes the proof of the main claim in Proposition 7.6: a boundary point of a surface is topologically different from an interior point.

Corollary 2 ('Fundamental theorem of algebra')   *Let*

$$f(z) = z^n + a_{n-1}z^{n-1} + \cdots + a_1 z + a_0$$

be a polynomial of degree $n \geq 1$ in z, with complex coefficients $a_i \in \mathbb{C}$. Then there exists a complex number ζ such that $f(\zeta) = 0$. In other words, $\mathbb{C}$ is algebraically closed.

Proof   Write $\mathbb{C}^* = \mathbb{C} \setminus \{0\}$. Obviously $\mathbb{C}^*$ is homeomorphic to D^*, so that the definition and properties of the winding number apply also to $\mathbb{C}^*$.

I first give the proof forgetting the small detail of the base point P_0, then explain how to patch this up. For $K \in \mathbb{R}$, $K \geq 0$, define

$$\varphi_K : [0, 1] \to \mathbb{C} \quad \text{by } t \mapsto f(K \exp(2\pi i t)).$$

If $\varphi_K(t) = 0$ for some K and some t then $f(\zeta) = 0$ for $\zeta = K \exp(2\pi i t)$. Assume by contradiction that this never happens. Then $\varphi_K : [0, 1] \to \mathbb{C}^*$ is a continuous family of loops in $\mathbb{C}^*$. When $K = 0$ it is the constant loop: $\varphi_0(t) = a_0$ for all t. When $K \gg 0$ it has winding number n. Indeed, if $K > 1 + \sum_{i=0}^{n-1} |a_i|$ the term z^n in $f(z)$ is bigger than all the other terms put together, so that the loop looks like $K^n(\sin nt + i \cos nt)$ plus a smaller error term that does not allow the path to reach to the origin.

However, by Theorem 7.15.3, if we assume that φ_K maps $[0, 1]$ to $\mathbb{C}^*$, the winding number must be constant, independent of K. This is a contradiction. Therefore, sometimes $f(z) = 0$.

The proof just given does not work as it stands, because Theorem 7.15.3 dealt only in based loops. There are several ways of dealing with this; one method would be to reprove Theorem 7.15.3 without base points, or to prove that the winding number does not depend on the choice of a base point.

An easy ad hoc method is to define a new family of paths φ_K starting from the base point $P_0 = a_0$ in the following way: we spend the first $1/3$ of the time in the interval $[0, 1]$ plodding out from $f(0) = a_0$ to $f(K) = \varphi_K(0)$ along the path $f(\mathbb{R})$; then we pursue the loop φ_K at 3 times the original speed, returning to $f(K) = \varphi_K(1)$ at time $t = 2/3$; then we spend the final $1/3$ of the time returning from $f(K)$ to $f(0)$ by retracing our steps along the same path $f(\mathbb{R})$. The new path has the same winding number as the old, because any change in the argument θ made in plodding out to $f(K)$ is exactly cancelled when we retraced our steps. The details are easy to work out.   QED

Exercises

7.1   Let (X, d) be a metric space. Check that Definition 7.2 does indeed define a topology on X; in other words, check that the set $\mathcal{T}$ of open sets in the metric sense is a topology. [Hint: use the triangle inequality.]

Questions on point-set topology.

7.2 X, Y, Z are topological spaces and $f: X \to Y$, $g: Y \to Z$ continuous maps. Prove that $g \circ f$ is continuous. Count the lines of your proof, and compare with the same proof in a standard analysis or metric spaces course.

7.3 X is a metric space with metric topology $\mathcal{T}_X$. Prove that a sequence of points $a_i \in X$ converges to l in the sense of the metric if and only if it converges in the sense of topology as in 7.4.2.

7.4   By definition, a sequence of points $\{x_i\}_{i=1,2,...}$ converges to $x \in X$ in a topological space if every neighbourhood U of x contains all but finitely many of the x_i. Let X, Y be topological spaces and $f: X \to Y$ continuous.

   (a) Prove that $\{x_i\}$ converge to x implies $\{f(x_i)\}$ converge to $f(x)$. That is, 'continuity implies sequential continuity' for topological spaces.

   (b) Conversely, prove that for a metric space X, this convergence for all sequences implies that f is continuous. In other words, 'sequential continuity implies continuity' for metric spaces.

   (c) Now let X be a topological space, not necessarily metric, in which every point $x \in X$ has a countable basis of neighbourhoods (referred to in 7.2). Prove sequential continuity implies continuity.

   (d) Prove that if X is an uncountable set with the cofinite topology (7.1 Example 2), then there does not exist a countable basis for the neighbourhoods of $x \in X$.

   (e) (Harder) Find a topological space and a map $f: Y \to X$ which is sequentially continuous but not continuous.

7.5 X is a metric space, $x, y \in X$ and $a_1, a_2, \ldots$ a sequence of points of X. Which of the following are topological properties?

(a) $X \setminus x$ is disconnected.

(b) $a_i \to x$ as $i \to \infty$.

(c) x is in the closure of $\{y\}$.

(d) a_i is a Cauchy sequence.

(e) The ball $B(x, 1)$ is compact.

(f) Every neighbourhood of x is a countable set.

(g) The closure of the ball $B(x, 1)$ is connected.

(h) For every compact subset $V \subset X$, the complement $X \setminus V$ is disconnected.

For each statement, give a proof or a counterexample, or both.

7.6   How many capital letters of the alphabet are there up to homeomorphism in a typeface without knobs on, such as

ABCDEFGHIJKLMNOPQRSTUVWXYZ?

Scrabble players do it with K and Q.

7.7 X and Y are topological spaces and $f : X \to Y$ a continuous surjective map. Prove that if X is sequentially compact, so is Y. [Hint: consider a sequence in Y and use the stated properties of f and X. Compare the proof of Proposition 7.4.2.]

7.8   Prove that a continuous function $f : X \to \mathbb{R}$ on a compact space X is bounded, and achieves its bounds. [Hint: to get bounded, just say balls, lots of balls, ... as before. Let $K = \sup f(X) \in \mathbb{R}$, which exists by the completeness axiom. By contradiction assume that $f(x) \neq K$ for all $x \in X$; consider the open sets $U_\varepsilon \subset X$ defined by $U_\varepsilon = \{x \mid f(x) \leq K - \varepsilon\}$.]

7.9   Prove that a continuous function $f : [a, b] \to \mathbb{R}$ is uniformly continuous. [Hint: for a given ε, the definition of continuity gives balls $B(x, \delta_x), \ldots$]

7.10 X is a topological space and $Y \subset X$ a subset with the subspace topology; prove that every *closed* subset of Y is of the form $Y \cap V$ with V closed in X.

7.11 X is a metric space and $Y \subset X$ a subset. Prove that the following two topologies on Y are identical.

(a) Take the metric topology $\mathcal{T}_X$ and the subspace topology $\mathcal{T}_{Y,1}$ on Y.

(b) Restrict the metric d_X to Y to get a metric d_Y, then take the metric topology $\mathcal{T}_{Y,2}$ on Y corresponding to d_Y.

7.12   Find all the possible topologies on a set $\{x, y\}$ with two points.

7.13   Study the possible topologies on a finite set.

(a) If a topological space is not T_1 (see 7.11) then there exist $x \neq y$ such that the constant sequence $y, y, \ldots$ converges to x. That is, x is in the closure of the set $\{y\}$.

(b) Write $x \, C \, y$ if x is in the closure of y, and think of this as a relation between x and y. Prove that C is a transitive relation.

(c) Define the relation $x \, R \, y$ by

$$x \, R \, y \iff x \, C \, y \text{ and } y \, C \, x.$$

Prove that R is an equivalence relation.

(d) Let $Y \subset X$ be an equivalence class of R; prove that the subspace topology on Y is the indiscrete topology (no opens other than $\emptyset$ and X).

(e) (Harder) Use steps (a)–(d) to describe all possible topologies on a finite set Y.

7.14 Let X be a topological space, $\sim$ an equivalence relation on X and $Y = X/\sim$ the quotient topological space. Think of the relation $\sim$ as the subset

$$Z(\sim) = \big\{(x, y) \mid x \sim y\big\} \subset X \times X$$

where $X \times X$ is given the product topology.

(a) By imitating the proof of Proposition 7.11, prove that Y is Hausdorff if and only if $Z(\sim) \subset X \times X$ is closed.

(b) Let $Z \subset X \times X$ be the *closure* of the diagonal, considered as a relation ($x \sim y$ if and only if $(x, y) \in Z$); describe what $x \sim y$ means in terms of neighbourhoods of x and y, and prove that $\sim$ is an equivalence relation.

(c) Prove that X has a continuous map $f : X \to X'$ to a Hausdorff space which has the UMP for such maps.

Exercises on surfaces.

7.15 Write down equations for a torus, a solid torus and a Möbius strip in terms of Cartesian coordinates (x, y, z) or cylindrical polar coordinates (r, θ, z) for $\mathbb{R}^3$. [Hint: you get a torus by rotating a circle about an axis outside it, and a Möbius strip by letting a diameter of the circle rotate simultaneously to get $1, 3, 5, \ldots$ half-twists.]

7.16 Prove that $S^2 \setminus \{2 \text{ points}\}$ is homeomorphic to the cylinder $S^1 \times \mathbb{R}$. [Hint: let the two points be the poles N and S, and think of Mercator's projection.]

7.17 Using Figure 7.7, prove the following statements.

(a) If $L = \mathbb{P}^1$ is the line obtained from the equatorial circle, then $\mathbb{P}^2 \setminus L$ is topologically a disc (the upper half-sphere), and a neighbourhood of L in $\mathbb{P}^2$ is a Möbius strip.

(b) If $Q = \{x^2 + y^2 = z^2\} \subset \mathbb{P}^2$ is a conic curve, then $\mathbb{P}^2 \setminus Q$ consists of two pieces, one a Möbius strip and the other a disc; a neighbourhood of Q in $\mathbb{P}^2$ is a cylinder.

Draw pictures illustrating the following statement: cutting $\mathbb{P}^2$ along a line is like cutting a Möbius strip along its central curve, whereas cutting $\mathbb{P}^2$ along a conic is like cutting a Möbius strip along the curve trisecting the width of the strip.

7.18 In 7.6, I obtained the Möbius strip, the cylinder and the torus from a square by glueing its edges in a particular fashion. In Figure 7.16a, I give two other glueing rules.

(a) Show that the first pattern builds a surface homeomorphic to the projective plane $\mathbb{P}^2$.

(b) Show that the second pattern corresponds to a surface that you can build in two steps, first glueing a cylinder as in Figure 7.6c and then identifying the circles at the ends, carefully remembering their orientation. This surface is called the *Klein bottle*. It shares with $\mathbb{P}^2$ the property that it cannot be embedded in $\mathbb{R}^3$ without self-crossing.

7.19 The top panel of Figure 7.16b shows a surface with two handles, with a set of circles marked on its surface, in analogy with the last panel of Figure 7.6c.

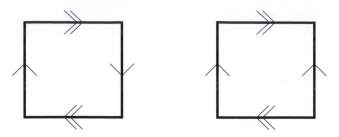

Figure 7.16a   Glueing patterns on the square.

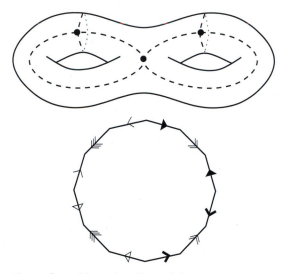

Figure 7.16b   The surface with two handles and the 12-gon.

(a) Verify that cutting the surface along the marked circles leads to the 12-gon on the bottom panel of Figure 7.16b, with the edges identified as shown. Hence conversely, glueing the 12-gon with the given pattern leads to a surface with two handles!

(b) Triangulate the surface by triangulating the 12-gon. Compute the Euler number 'faces − edges + vertexes'. Compare 9.4.

Exercises on loops

7.20   Draw the graph of the function

$$f^{(s)}(t) = \begin{cases} 4t/s & \text{for } 0 \le t \le s/4 \\ 2 - 4t/s & \text{for } s/4 \le t \le s/2 \\ 0 & \text{for } s/2 \le t \le 1. \end{cases}$$

Here $s \in (0, 1]$. Have you seen anything like this before? Set $f^{(0)}(t) = 0$, and prove the following:

(a) for any fixed $s \in [0, 1]$ the formula $\varphi^{(s)}(t) = (f^{(s)}(t), t)$ defines a path $\varphi^{(s)} \colon [0, 1] \to \mathbb{R}^2$ (i.e. it is continuous);

(b) for fixed $t \in [0, 1]$ the map $s \mapsto \varphi^{(s)}(t)$ is continuous;

(c) $\varphi^{(s)}$ is not a continuous family of paths in $\mathbb{R}^2$ in the sense of Definition 7.15.1;

(d) $\varphi^{(s)}$ is something you would not do to a dog lead;

(e) $\Phi(s, t) = f^{(s)}(t)$ is not a continuous function of s, t near $(0, 0)$.

The point of the question is to justify the tentative definition in 7.15.1, in particular to convince you of the requirement for uniformity in t.

7.21   Suppose that X is a path connected topological space and pick two points $P_0, Q_0 \in X$. Prove that all loops in X based at P_0 are contractible if and only if all loops in X based at Q_0 are contractible. [Hint: compare the end of 7.15.4.]

8 Quaternions, rotations and the geometry of transformation groups

Chapters 1–5 discussed transformations that depend continuously on parameters: for example, Euclidean rotations in the plane that depend on the centre and the angle of rotation. I stressed that composition of transformations is a natural operation, an idea that led in Chapter 6 to the definition of a geometric transformation group. Here I focus on groups with a continuous family of elements, especially some examples arising in geometry where the group of transformations has an interesting geometry of its own. The discussion is a first introduction to some of the basic ideas of 'continuous transformation groups'. The formal definition and a detailed treatment of this type of 'group-manifold' (or *Lie group*) is beyond the scope of this book, but see 8.8 and Segal [22].

As an example, recall that a rotation of $\mathbb{E}^2$ around a fixed point P is given by the matrix $\left(\begin{smallmatrix} \cos\theta & -\sin\theta \\ \sin\theta & \cos\theta \end{smallmatrix} \right)$, and so depends continuously on the real parameter θ. This parameter takes values in a circle. Thus the group of rotations of $\mathbb{E}^2$ around a fixed point has a geometry of its own, that of the circle, as shown in Figure 8.0. The relation between rotations in the plane and the circle can be conveniently expressed in terms of complex numbers, with the action of rotation by θ on the column vector $\left(\begin{smallmatrix} x \\ y \end{smallmatrix} \right)$ written as multiplication of the complex number $x + iy$ by the complex number $\exp(i\theta)$ of absolute value 1. On the other hand, the set of unit complex numbers is the circle S^1 in the complex plane.

A highlight of this chapter is Corollary 8.5.3, which applies the homeomorphism criterion Theorem 7.14 (one of the main results of Chapter 7) to give a description in similar terms of the topology of the groups of rotations of $\mathbb{E}^3$ and $\mathbb{E}^4$ around a fixed point. The algebra of complex numbers is replaced by the algebra of *quaternions*

$$\mathbb{H} = \left\{ a + bi + cj + dk \right\} \quad \text{with } a, b, c, d \in \mathbb{R},$$

where i, j, k all square to -1 and multiply together wisely. Corollary 8.5.3 describes the topology of the group of three- and four-dimensional rotations in terms of the sphere S^3 of unit quaternions.

The group of three dimensional rotations is of basic importance in many areas of mechanics and physics, describing symmetries of Euclidean space $\mathbb{E}^3$, a space that old-fashioned empiricists believe we inhabit. The quantum mechanical treatment of

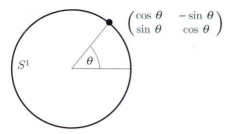

Figure 8.0   The geometry of the group of planar rotations.

the *spin* of the electron is a pretty illustration of my treatment of the topology of the group of three-dimensional rotations. As most ingredients are at hand already, I cannot resist the temptation to include a section on this, cribbed more or less directly from Feynman [7]. The discussion puts together in a very satisfactory way ideas from algebra (groups, algebra of quaternions), analysis (topology, compactness), geometry (rotations of $\mathbb{E}^3$) and quantum physics (wave function, spin of the electron).

8.1 Topology on groups

A group G is a *topological group* if it has a topology defined on it so that multiplication and inverse are continuous. In more detail, a topological group is an object G having two quite different structures: a collection of open subsets satisfying the axioms for a topology, and a multiplication map with identity and inverse satisfying the group axioms. I require the group structure to respect the topological structure in the sense that

$$\text{mult}\colon\ G \times G \to G \quad \text{and} \quad \text{inv}\colon\ G \to G$$
$$(g, h) \mapsto gh \qquad\qquad g \mapsto g^{-1}$$

are both continuous maps of topological spaces; here $G \times G$ has the product topology of 7.10.

Example 1   Any finite group G is a topological group under the discrete topology.

Example 2   The groups $(\mathbb{R}, +)$ and $(\mathbb{R}^*, \times)$ are topological groups with respect to the usual topology of $\mathbb{R}$. This is just a fancy way of restating the fact, used all over the place in a first analysis course, that the four operations addition, subtraction, multiplication and division are continuous on the reals.

Example 3   A substantial generalisation of the previous example brings us back to the linear geometries of Chapters 1–5. Recall the *general linear group* $GL(n, \mathbb{R})$ of $n \times n$ real invertible matrixes. Note that $GL(n, \mathbb{R})$ is a subset of the set of real matrixes $M(n \times n, \mathbb{R}) = \mathbb{R}^{n^2}$. This latter is a metric space, and therefore has a natural metric topology. Moreover, it is an easy fact that matrix multiplication and inverse are continuous. Hence $GL(n, \mathbb{R})$ is a topological group. As a consequence, affine transformations $Aff(n, \mathbb{R})$ (compare 4.5) also form a topological group.

The group $\mathbb{R}^*$ of constant diagonal matrixes is a subgroup of $GL(n + 1, \mathbb{R})$, the *centre* of $GL(n + 1, \mathbb{R})$, that is, the subgroup of elements commuting with every element $g \in GL(n + 1, \mathbb{R})$ (see 5.5 and Exercise 6.3). The quotient

$$PGL(n + 1, \mathbb{R}) = GL(n + 1, \mathbb{R})/\mathbb{R}^*$$

is a topological group with the quotient topology. This is of course the group of projective linear transformations of $\mathbb{P}^n$ familiar from 5.5, the *projective linear group*.

Example 4   The *orthogonal group*

$$O(n) = \left\{ A \in GL(n, \mathbb{R}) \mid {}^tAA = 1_n \right\},$$

the group of orthogonal $n \times n$ matrixes, is a topological group in the subspace topology. Hence also $Eucl(n)$, the group of Euclidean motions, and the group of motions of S^2 (see 3.5) are topological groups.

Example 5   Hyperbolic motions form a matrix group, the *Lorentz group* or group of Lorentz transformations (see 3.11 for the notation and compare Theorem 3.11 and Exercise 8.5)

$$O^+(1, 2) = \left\{ A \in GL(3, \mathbb{R}) \,\middle|\, \begin{matrix} {}^tAJA = J, \text{ and } A \text{ preserves the} \\ \text{halves of the cone } q_L(\mathbf{v}) < 0 \end{matrix} \right\}.$$

This is also a topological group. It and its higher dimensional colleagues $O^+(1, n)$ are important in special relativity and related areas of physics.

The topological groups in Examples 2–5 have an interesting 'continuous' geometry. Here is a simple example (see Figure 8.0): recall that $O(2)$ is the group of all rotation matrixes $\left(\begin{smallmatrix} \cos\theta & -\sin\theta \\ \sin\theta & \cos\theta \end{smallmatrix} \right)$ and reflection matrixes $\left(\begin{smallmatrix} \cos\theta & \sin\theta \\ \sin\theta & -\cos\theta \end{smallmatrix} \right)$. Thus $O(2)$ is a union of two connected components, each a copy of the circle S^1 parametrised by the angle θ. One aim of this chapter is to generalise this nice description to some other orthogonal groups.

8.2   Dimension counting

Here I begin the study of some particular aspects of the geometry of transformation groups. In this section I want to concentrate on a measure of their size. Recall that $O(2)$ can be described geometrically as the union of two circles. The circle S^1 is a one dimensional geometric object in the sense that its points depend on one real parameter θ; standing at a point of the circle, there is one direction in which you can move.

Without going into rigorous details, by *dimension* of a transformation group G, denoted $\dim G$, I understand the number of continuous real parameters needed to characterise an element $g \in G$. The previous paragraph then shows that dim $O(2) = 1$. Do not get confused by the fact that $O(2)$ has two components; to characterise elements of $O(2)$, I need one continuous real parameter (the angle θ) and a discrete parameter (the choice of one of the components, equivalently the sign of the determinant, or its value ± 1).

I proceed to compute the dimension of transformation groups in some nontrivial cases. The computations will be performed by describing elements of the groups in a way which makes it possible to count the parameters involved directly.

Proposition   *An element $g \in \mathrm{Eucl}(n)$ depends on $\binom{n+1}{2}$ real parameters, so* dim $\mathrm{Eucl}(n) = \binom{n+1}{2}$. *Further,*

$$\dim O(n) = \binom{n}{2}, \quad \dim \mathrm{GL}(n, \mathbb{R}) = n^2, \quad \dim \mathrm{PGL}(n+1, \mathbb{R}) = n(n+2).$$

Proof   The language of Euclidean frames from 1.12 gives a way of specifying elements of the Euclidean group. Choose a reference frame $\{P_0, P_1, \ldots, P_n\}$; then by Theorem 1.12, elements of the Euclidean group $\mathrm{Eucl}(n)$ correspond one-to-one with the set of Euclidean frames $\{Q_0, Q_1, \ldots, Q_n\}$. Now calculate:

- $Q_0 \in \mathbb{E}^n$ is any point, so depends on n parameters;
- $Q_1 \in \mathbb{E}^n$ is any point with $d(Q_0, Q_1) = 1$, that is, it is any point of the unit sphere S^{n-1} with centre Q_0, hence depends on $n - 1$ real parameters;
- writing $\mathbf{e}_1 = \overrightarrow{P_0 P_1}$ and $\mathbf{e}_1^\perp = \mathbb{E}^{n-1} \subset \mathbb{E}^n$ for the orthogonal complement, Q_2 is given by a point of the unit sphere $S^{n-2} \subset \mathbb{E}^{n-1}$, so depends on $n - 2$ real parameters;
- similarly, Q_i is given by a point of S^{n-i}, and hence depends on $n - i$ real parameters;
- in particular, Q_n is one of two points, so has no continuous parameter.

Thus a Euclidean frame depends on

$$\dim \mathrm{Eucl}(n) = n + (n - 1) + \cdots + 1 + 0 = \binom{n + 1}{2}$$

parameters.

An element of $O(n)$ fixes the origin, which I can take to be $P_0 = Q_0$ in the above argument. Hence the dimension count is

$$\dim O(n) = (n - 1) + \cdots + 1 + 0 = \binom{n}{2},$$

agreeing with dim $O(2) = 1$. Said slightly differently, $O(n)$ and $\mathrm{Eucl}(n)$ differ by the translation part (compare Proposition 6.5.3), which accounts for n parameters:

$$\dim O(n) = \dim \mathrm{Eucl}(n) - n = \binom{n + 1}{2} - n = \binom{n}{2}.$$

The dimension of the general linear group can be calculated in exactly the same way. Elements of $\mathrm{GL}(n, \mathbb{R})$ correspond to invertible maps of the vector space $\mathbb{R}^n$. Such a map is determined by the images of the n usual basis vectors in $\mathbb{R}^n$, parametrised by a total of n^2 numbers (the entries of the matrix representing the map). Not all parametrisations give invertible maps, but most do: I only have to exclude matrixes with zero determinant. Hence there are n^2 real parameters involved, so

$$\dim \mathrm{GL}(n, \mathbb{R}) = n^2.$$

Finally by Theorem 5.5 there are as many projective transformations as projective frames of reference. Hence I have to pick $n + 2$ general points in $\mathbb{P}^n$, leading to

$$\dim \mathrm{PGL}(n + 1, \mathbb{R}) = (n + 2)n$$

parameters. Incidentally, the dimension of the projective group can also be calculated from its definition $\mathrm{PGL}(n + 1, \mathbb{R}) = \mathrm{GL}(n + 1, \mathbb{R})/\mathbb{R}^*$, which gives

$$\dim \mathrm{PGL}(n + 1, \mathbb{R}) = \dim \mathrm{GL}(n + 1, \mathbb{R}) - 1$$
$$= (n + 1)^2 - 1 = (n + 2)n. \quad \text{QED}$$

You can design your own parameter counts for some other groups not mentioned in the proposition; for example, do and generalise Exercise 8.3.

8.3   Compact and noncompact groups

Proposition   *The orthogonal group* $\mathrm{O}(n)$ *is a compact topological space.*

Proof   This is a simple application of Proposition 7.4.2. The orthogonal group is a matrix group: it is a subspace of the space $\mathbb{R}^{n^2}$ of real matrixes. Hence it is enough to show that it is closed and bounded. The equation ${}^t\!A A = 1_n$ defines a closed subset of $\mathbb{R}^{n^2}$, so the main issue is boundedness. However, if $A = (a_{ij})$ is orthogonal, then its columns form an orthonormal basis and in particular for every $1 \le k \le n$, $\sum_{i=1}^n a_{ik}^2 = 1$. Hence

$$\sum_{i,k=1}^n a_{ki}^2 = n$$

which just says that every orthogonal matrix A is contained in a ball of radius $\sqrt{n}$ in $\mathbb{R}^{n^2}$.   QED

A compact space is often much more pleasant to work with than a noncompact one. However, many transformation groups are visibly noncompact, such as the additive group $\mathbb{R}$. On the other hand, the topology and geometry of $\mathbb{R}$ are very simple (for example, $\mathbb{R}$ is simply connected, and can be parametrised by a real parameter without overlap). Most transformation groups are of course more complicated; however, in a suitable sense they can be *topologically decomposed* as a compact group times a group homeomorphic to $\mathbb{R}^n$.

Example 1   The simplest example is the multiplicative group $\mathbb{R}^*$ of nonzero real numbers. There is a homeomorphism (in this case, an isomorphism of groups)

$$\mathbb{R}_+ \times \{\pm 1\} \to \mathbb{R}^*;$$

in plain English, every nonzero number is the product of a positive number and a sign. The space $\mathbb{R}_+$ is homeomorphic to $\mathbb{R}$; the group $\{\pm 1\}$ is finite so clearly compact.

Example 2   Although the next example looks similarly innocent, it appears in many different guises throughout geometry, Fourier analysis, Lie groups, representation theory, complex analysis and number theory. Consider the multiplicative group $\mathbb{C}^*$ of nonzero *complex* numbers. This is a topological group; for example, I can view $\mathbb{C}$ as the plane $\mathbb{R}^2$ and take the subspace topology. The space $\mathbb{C}^*$ is obviously noncompact. However, there is a homeomorphism (even a group isomorphism)

$$\begin{aligned} S^1 \times \mathbb{R}_+ &\to & \mathbb{C}^* \\ (\theta, r) &\mapsto & r\exp(i\theta). \end{aligned}$$

Here S^1 is compact (and definitely not homeomorphic to a product of copies of $\mathbb{R}$, which is the essential content of 7.15.4, Corollary 1) and $\mathbb{R}_+$ is homeomorphic to $\mathbb{R}$.

Example 3   The final example is more substantial, and deals with the difference between the groups $GL(n, \mathbb{R})$ and $O(n)$. Write $T_+(n) \subset GL(n, \mathbb{R})$ for the set of upper triangular matrixes with positive diagonal entries:

$$T_+(n) = \left\{ M = (m_{ij}) \in GL(n, \mathbb{R}) \mid m_{ij} = 0 \text{ for all } i > j, \text{ and } m_{ii} > 0 \right\}$$

$$= \left\{ \begin{pmatrix} + & * & \cdots & \\ 0 & + & * & \cdots \\ 0 & \cdots & \ddots & * \\ 0 & \cdots & 0 & + \end{pmatrix} \right\}.$$

It is easy to see that $T_+(n) \subset GL(n, \mathbb{R})$ is a subgroup.

Theorem   *Every element $A \in GL(n, \mathbb{R})$ can be written in a unique way in the form $A = BC$, where $B \in O(n)$ is an orthogonal matrix and $C \in T_+(n)$ is an upper triangular matrix with positive diagonal entries. Moreover, B and C depend continuously on A. The map*

$$GL(n, \mathbb{R}) \to O(n) \times T_+(n) \quad \textit{given by} \quad A \mapsto (B, C)$$

is a homeomorphism (see 7.3, but not a group homomorphism!).

Discussion   The space $O(n)$ is compact by the above Proposition. The space $T_+(n)$ is homeomorphic to $\mathbb{R}^N$, where $N = \binom{n+1}{2}$. Many geometric questions on $GL(n, \mathbb{R})$

reduce to similar questions on O(n); for a simple example, compare Remark 8.4. Note also the dimension count:

$$\dim O(n) + \dim T_+(n) = \binom{n}{2} + \binom{n+1}{2} = n^2 = \dim \mathrm{GL}(n, \mathbb{R}).$$

Proof   I view the $n \times n$ matrix A as a row made up of n column vectors $\mathbf{f}_i$. Thus $\{\mathbf{f}_1, \dots, \mathbf{f}_n\}$ is a basis of $\mathbb{R}^n$ because $A \in \mathrm{GL}(n, \mathbb{R})$. If it is an orthonormal basis then there is no problem: $A \in O(n)$, and we must take $B = A$ and $C = 1$. If A is not orthogonal to start with, then the Gram–Schmidt process described in the proof of Theorem B.3 (1) produces an orthonormal basis. Set B to be the matrix formed from the new basis vectors as columns, and C to be the matrix describing the change of basis. Clearly $B \in O(n)$; I leave you to check (see Exercise 8.6) that $C \in T_+(n)$ and that B, C depend continuously on A. Then the map $A \mapsto (B, C)$ is continuous, and its inverse is matrix multiplication $(B, C) \mapsto BC$.   QED

8.4   Components

Recall from 7.4.1 that every topological space can be decomposed into a number of components, which are themselves connected. I repeatedly discussed the geometry of O(2): a union of two circles. A circle S^1 is connected, so O(2) has two connected components. This is typical:

Proposition   *The group* O(n) *has two connected components, distinguished by* $\det A = \pm 1$.

Remark   One can use Theorem 8.3 to show that GL(n, $\mathbb{R}$) also has two connected components, that are distinguished by $\det A > 0$ and $\det A < 0$; see Exercise 8.4. The group O(1, 2) of all Lorentz matrixes has 4 components, as discussed in Exercise 8.5.

Proof   An orthogonal matrix has determinant ± 1. (Compare 1.10; recall that I called A direct if $\det A = 1$ and opposite if $\det A = -1$.) The function

$$\det \colon O(n) \to \{\pm 1\}$$

is continuous, so the two possibilities $\det A = \pm 1$ determine two disjoint open and closed sets of O(n). It remains to show that each of these sets is path connected.

Fix a matrix $A \in O(n)$. By the normal form theorem 1.11, A can be written with respect to a suitable orthonormal basis in the diagonal block form with 2×2 diagonal blocks

$$B_i = \begin{pmatrix} \cos \theta_i & -\sin \theta_i \\ \sin \theta_i & \cos \theta_i \end{pmatrix},$$

and one optional block ± 1. For t varying from 0 to 1, let $A(t)$ be the matrix with the same block form as A, but with blocks

$$B_i(t) = \begin{pmatrix} \cos t\theta_i & -\sin t\theta_i \\ \sin t\theta_i & \cos t\theta_i \end{pmatrix}.$$

The rule $t \mapsto A(t)$ gives a continuous path $[0, 1] \to$ O(n) joining A either to the identity or to the element diag$(1, \ldots, 1, -1)$. Therefore, the two subsets of O(n) defined by det $A = \pm 1$ are both path connected. A path connected space is connected by Lemma 7.4.1 (2). QED

The *special orthogonal group* is the group

$$\text{SO}(n) = \{A \in \text{O}(n) \mid \det A = 1\}.$$

By the Proposition, this is a connected component of O(n). Since it is the kernel of a group homomorphism det: O(n) $\to \{\pm 1\}$, it is also a normal subgroup of index 2 in O(n).

In the special case $n = 3$, the elements of SO(3) can be described explicitly. By the normal form theorem 1.11, any orthogonal 3×3 matrix of determinant 1 has the form

$$\begin{pmatrix} 1 & & \\ & \cos\theta & -\sin\theta \\ & \sin\theta & \cos\theta \end{pmatrix}$$

in a suitable basis. If l is the line through the origin with direction vector given by the first basis element, then the motion of $\mathbb{E}^3$ described by this matrix is the rotation Rot(l, θ) around the line l. Hence SO(3) is the *group of rotations of* $\mathbb{E}^3$ *about axes passing through O.*

8.5 Quaternions, rotations and the geometry of SO(n)

As I discussed before, for $n = 2$ the group SO(2) is homeomorphic to the circle S^1. The purpose here is to find a similar description of the special orthogonal groups SO(3) and SO(4) in terms of the 3-sphere. I start with a small detour to introduce the quaternions, the main protagonists in the game. Note that SO(n) is the group of direct motions of $\mathbb{E}^n$ with a fixed point, or in other words the group of rotations of $\mathbb{E}^n$; hence the aim is to find a connection between quaternions and rotations (for $n = 3, 4$).

8.5.1

Quaternions

The *algebra of quaternions* is the real vector space

$$\mathbb{H} = \{a + bi + cj + dk\} \quad \text{with } a, b, c, d \in \mathbb{R},$$

with the multiplication law

$$i^2 = j^2 = k^2 = -1, \quad ij = k, jk = i, ki = j, \quad ji = -k, kj = -i, ik = -j.$$

The cyclic symmetry makes this easy to remember.

Some terminology, similar to the traditional language of complex numbers: if $q = a + bi + cj + dk$, write $q^* = a - bi - cj - dk$ for the *conjugate* quaternion. We say that q is *real* if $b = c = d = 0$ and *pure imaginary* if $a = 0$.

Proposition

(1) $\mathbb{H}$ is an associative noncommutative $\mathbb{R}$-algebra of dimension 4 over $\mathbb{R}$.

(2)   The conjugation $q \mapsto q^*$ is an antiinvolution, *meaning*

$$(pq)^* = q^* p^* \quad \text{for all } p, q \in \mathbb{H}.$$

(3) $|q|^2 = qq^* = q^*q = a^2 + b^2 + c^2 + d^2$ is a positive definite quadratic form on $\mathbb{H}$; therefore for any nonzero $q \in \mathbb{H}$, the element

$$q^{-1} = q^*/|q|^2$$

is a 2-sided inverse of q. Hence $\mathbb{H}$ is a division algebra *or* skew field.

(4)   If $q \in \mathbb{H}$ and $q \notin \mathbb{R}$, then $q = A + BI$ with I pure imaginary, $I^2 = -1$ and $A, B \in \mathbb{R}$. Hence the subalgebra $\mathbb{R}[q]$ of $\mathbb{H}$ generated by q is of the form $\mathbb{R}[q] \cong \mathbb{C} \subset \mathbb{H}$.

(5)   If I is pure imaginary with $I^2 = -1$, there exists $J, K \in H$ such that I, J, K have the same multiplication table as i, j, k, that is $I^2 = J^2 = K^2 = -1$ and $IJ = K$, etc.

Proof   (1) Noncommutativity is clear from the multiplication table: $ij = k \neq -k = ji$.

Because everything is $\mathbb{R}$-linear, it is enough to check the associative law $a(bc) = (ab)c$ for the basis elements $a, b, c \in \{1, i, j, k\}$. If any of a, b, c is 1 then it is OK. By the cyclic symmetry, I can assume that the first term $a = i$; if only i appears, then I am working in a copy of $\mathbb{C}$. This leaves only 8 cases to check by brute force:

$$
\begin{aligned}
i(ij) &= ik = -j = (i^2)j; &\quad i(ik) &= i(-j) = -k = (i^2)k; \\
i(ji) &= i(-k) = j = ki = (ij)i; &\quad i(j^2) &= -i = kj = (ij)j; \\
i(jk) &= i^2 = -1 = k^2 = (ij)k; &\quad i(ki) &= ij = k = -ji = (ik)i; \\
i(kj) &= -i^2 = 1 = -j^2 = (ik)j; &\quad i(k^2) &= -i = -jk = (ik)k.
\end{aligned}
$$

This is of course pure gobbledygook. A much more convincing argument is to say that i, j, k are maps of something, such that multiplication coincides with composition of maps, so is associative for a fundamental reason; see Exercise 8.8.

(2) Again because everything is $\mathbb{R}$-linear, it is enough to check that $(pq)^* = q^* p^*$ for basis elements $a, b \in \{1, i, j, k\}$. The brute force method is an easy exercise: $(1i)^* = -i = (i^*)(1^*)$, $(ij)^* = -k = (-j)(-i)$, etc.; see Exercise 8.9.

(3) On multiplying out the product $(a + bi + cj + dk)(a - bi - cj - dk)$, the terms $a^2 + b^2 + c^2 + d^2$ appear in the obvious way from the squared terms. The cross terms all cancel out, either as $(a \times -bi) + (bi \times a) = 0$ or $(bi \times -cj) + (cj \times -bi) = -bc(i \times j + j \times i) = 0$.

(4) Note that $q + q^* = 2a$ and $qq^* = |q|^2 \in \mathbb{R}$, so that q and q^* are the two roots of a quadratic polynomial $x^2 - 2ax + |q|^2$ with real coefficients. Also, $q - q^* = 2(bi + cj + dk)$ is pure imaginary, and an easy calculation similar to that in (3) shows that $(q - q^*)^2 = -4(b^2 + c^2 + d^2) < 0$ (because $q \notin \mathbb{R}$), so that this has no real roots. Thus $q = A + BI$ where $A = a$, $B = \sqrt{(b^2 + c^2 + d^2)}$ and I is pure imaginary with $I^2 = -1$.

(5) is worked out as an exercise in Exercise 8.12.   QED

Remark   (3) says that the Euclidean distance on $\mathbb{R}^4 = \mathbb{H}$ is determined by the algebra structure of $\mathbb{H}$ together with the antiinvolution $q \mapsto q^*$. This has various nice corollaries. For example, the direct sum decomposition

$$\mathbb{H} = \{\text{real quaternions}\} \oplus \{\text{imaginary quaternions}\} = \mathbb{R} \oplus \mathbb{R}^3$$

is orthogonal. Also, two imaginary vectors p, q anticommute $pq = -pq$ if and only if the corresponding vectors of $\mathbb{R}^3$ are orthogonal. This point is the main reason that quaternions can be applied to rotations of $\mathbb{E}^3$ and $\mathbb{E}^4$.

8.5.2
Quaternions
and
rotations

Set

$$U = \{\text{unit quaternions}\} = \{q \in \mathbb{H} \mid qq^* = 1\} = S^3 \subset \mathbb{R}^4$$

for the unit quaternions. Note that U has two structures: it is a group under multiplication, and also has its own geometry as the sphere S^3. The two structures are compatible as in 8.1. The group U generalises the multiplicative group of complex numbers of modulus 1, which is the unit circle $S^1 \subset \mathbb{C}$.

For the next theorem, identify $\mathbb{H}$ and its quadratic form $|q|$ with $\mathbb{E}^4$ and its Euclidean distance. The purely imaginary quaternions form a linear subspace which gets identified with $\mathbb{E}^3$.

Theorem

(1)   For any $p \in U$, left multiplication $a_p \colon x \mapsto px$ defines a map $\mathbb{H} \to \mathbb{H}$ which is a direct motion of $\mathbb{H} = \mathbb{E}^4$ fixing the origin; the same holds for right multiplication $b_q \colon x \mapsto xq^*$.

(2)   The group homomorphism $\varphi \colon U \times U \to SO(4)$ defined by

$$\varphi(p, q) = a_p \circ b_q \colon x \mapsto pxq^*$$

is surjective, and $\varphi(p, q) = \mathrm{id}_\mathbb{H}$ if and only if $(p, q) = (1, 1)$ or $(p, q) = (-1, -1)$.

(3)   For any $q \in U$, the map $r_q \colon x \mapsto qxq^*$ is a direct motion of $\mathbb{H} = \mathbb{E}^4$, which is the identity on real elements of $\mathbb{H}$ and takes pure imaginary quaternions of $\mathbb{H}$ to pure imaginary quaternions. Thus it defines a rotation of the subspace $\mathbb{E}^3 \subset \mathbb{H}$ of pure imaginary quaternions.

(4)   Any $q \in U$ with $q \notin \mathbb{R}$ has a unique expression in the form $q = \cos\theta + I\sin\theta$, where $I \in U$ is a pure imaginary quaternion and $\theta \in (0, \pi)$. Then $r_q = \mathrm{Rot}(I, 2\theta)$ is the rotation of $\mathbb{R}^3$ about the directed axis defined by I through the angle 2θ.

(5)   The group homomorphism $\psi \colon U = S^3 \to SO(3)$ defined by

$$\psi(q) = r_q$$

is surjective, and $\psi(q_1) = \psi(q_2)$ if and only if $q_1 = \pm q_2$.

Proof (1) It is clear that a_p is a motion, since it fixes 0 and $|px|^2 = |x|^2$. Moreover, it must be a direct motion, for example, because $\det(a_q)$ is a continuous map from the connected set $U = S^3$ to ± 1. (Several other proofs are possible, see Exercise 8.15.)

I relegate (2) to Exercise 8.22.

(3) is obvious, since $a \in \mathbb{R}$ commutes with quaternion multiplication, so $r_q(a) = qaq^* = aqq^* = a$. Also, if $p^* = -p$, then $r_q(p) = qpq^*$ has $(r_q(p))^* = (qpq^*)^* = qp^*q^* = -qpq^*$, so qpq^* is pure imaginary.

(4) follows from Proposition 8.5.1 (4): $\mathbb{R}[q] \cong \mathbb{C}$. The equation $x^2 = -1$ has exactly two roots $\pm I$ in $\mathbb{C}$, and choosing the appropriate sign gives $q = \cos\theta + I\sin\theta$ with $\theta \in (0, \pi)$. Then $r_q(I) = I$ follows because $\mathbb{R}[q] \cong \mathbb{C}$, so that $q^* = q^{-1}$ and $qIq^{-1} = I$.

Now let J, K be as in Proposition 8.5.1 (5). Then

$$qJq^* = (\cos\theta + I\sin\theta)J(\cos\theta - I\sin\theta)$$
$$= (\cos^2\theta - \sin^2\theta)J + (2\sin\theta\cos\theta)K,$$

and similarly $qKq^* = -(2\sin\theta\cos\theta)J + (\cos^2\theta - \sin^2\theta)K$. Thus r_q fixes the directed axis defined by I, and performs a rotation by 2θ in the plane spanned by J, K.

Finally (5) follows by (4); every rotation is hit exactly twice because of the 2θ. **QED**

**8.5.3
Spheres and
special
orthogonal
groups**

After all this algebra, come the relations between groups of rotations and the sphere S^3.

Corollary

(1) *There is a homeomorphism*

$$SO(3) \simeq S^3/\sim,$$

where $\sim$ is the equivalence relation on S^3 that identifies antipodal points $\mathbf{x}$ and $-\mathbf{x}$.

(2) *There is a homeomorphism*

$$SO(4) \simeq (S^3 \times S^3)/\approx,$$

where $\approx$ is the equivalence relation on $S^3 \times S^3$ that identifies $(\mathbf{x}, \mathbf{y})$ with $(-\mathbf{x}, -\mathbf{y})$.

Proof Both statements are direct corollaries of the previous theorem together with Theorem 7.14 and the definition of the quotient topology and its UMP discussed in 7.5.

In more detail, by Theorem 8.5.2 (5) there is a continuous surjective map $\psi\colon S^3 \to SO(3)$, with $\psi(\mathbf{x}) = \psi(\mathbf{y})$ if and only if $\mathbf{x} = \mathbf{y}$ or $\mathbf{x} = -\mathbf{y}$. By the universal mapping property 7.5 of the quotient topology, there is consequently a continuous map $\overline{\psi}\colon (S^3/\sim) \to SO(3)$ that is clearly a bijection. Now S^3 is compact, and therefore so is $S^3/\sim$ by Proposition 7.4.3. Also the subspace topology of $SO(3) \subset \mathbb{R}^9 = \{3 \times 3 \text{ matrixes}\}$ is metric and therefore Hausdorff. Therefore all the assumptions of Theorem 7.14 are satisfied, $\overline{\psi}$ is a homeomorphism, and (1) follows. (2) is proved in exactly the same way using the map $\varphi \colon U \times U \to \mathrm{SO}(4)$ of Theorem 8.5.2 (2).   QED

Remark   The statements of the corollary generalise for all n; namely, there exists a compact topological group $\mathrm{Spin}(n)$ called the *spinor group* with a surjective homomorphism $\pi \colon \mathrm{Spin}(n) \to \mathrm{SO}(n)$ with kernel $\langle \iota \rangle$ of order 2, so that π induces an isomorphism of groups $\mathrm{Spin}(n)/\langle \iota \rangle \to \mathrm{SO}(n)$ that is also a homeomorphism [15]. The pleasant thing about low dimensions is the fact that the spinor groups are spheres or products of spheres: $\mathrm{Spin}(2) \simeq S^1$, $\mathrm{Spin}(3) \simeq S^3$, $\mathrm{Spin}(4) \simeq S^3 \times S^3$.

8.6 The group SU(2)

In this brief section, I identify the group U of unit quaternions of 8.5 as a matrix group. This involves more linear algebra over the complex numbers, a subject that already made a brief but important appearance in 1.11.

Let V be a 2-dimensional $\mathbb{C}$-vector space together with a positive definite Hermitian form, represented in some basis by $|z_1|^2 + |z_2|^2$, or the matrix $\left(\begin{smallmatrix} 1 & 0 \\ 0 & 1 \end{smallmatrix}\right)$ (see B.6 for more details on Hermitian forms). A complex linear transformation of V that preserves this form is *unitary*: thus a matrix $A \in \mathrm{GL}(2, \mathbb{C})$ is unitary if it satisfies ${}^{\mathrm{h}}AA = I_n$, where ${}^{\mathrm{h}}A$ is the Hermitian conjugate defined by $({}^{\mathrm{h}}A)_{ij} = \overline{A}_{ji}$. The group of all such matrixes is the *unitary group* $\mathrm{U}(2)$. I am interested in its subgroup, the *special unitary group*

$$\mathrm{SU}(2) = \big\{ A \in \mathrm{U}(2) \,\big|\, \det A = 1 \big\}.$$

As matrix groups, both $\mathrm{U}(2)$ and $\mathrm{SU}(2)$ are topological groups in an obvious way.

Remark   A unitary matrix A has $|\det A| = 1$; see Exercise B.4. Thus the set of possible values for the determinant is the unit circle S^1, which is connected. Thus $\mathrm{SU}(2)$ is a normal subgroup, but not a connected component of $\mathrm{U}(2)$ in the same way as $\mathrm{SO}(2)$ is in $\mathrm{O}(2)$.

I write out explicitly the condition for a matrix $A \in \mathrm{GL}(2, \mathbb{C})$ to be special unitary (compare 1.11.1). If $A = \left(\begin{smallmatrix} a & b \\ c & d \end{smallmatrix}\right)$, the equations are

$$\begin{aligned} a\overline{a} + c\overline{c} &= 1, \\ a\overline{b} + c\overline{d} &= 0, \quad \text{and} \quad \det A = ad - bc = 1. \\ b\overline{b} + d\overline{d} &= 1, \end{aligned} \qquad (1)$$

One solves these equations more-or-less as in 1.11.1 to get $d = \overline{a}$ and $c = -\overline{b}$, where $a\overline{a} + b\overline{b} = 1$; see Exercise 8.20. Thus

$$\mathrm{SU}(2) = \left\{ \begin{pmatrix} a & b \\ -\overline{b} & \overline{a} \end{pmatrix} \,\middle|\, a, b \in \mathbb{C},\ |a|^2 + |b|^2 = 1 \right\}.$$

This description has an important corollary.

Corollary   *The map* $\left(\begin{smallmatrix} a & b \\ -\bar{b} & \bar{a} \end{smallmatrix}\right) \mapsto a + bj$ *defines an isomorphism from* SU(2) *to the group U of unit quaternions of 8.5.2.*

Proof   Write $a = a_1 + a_2 i$ and $b = b_1 + b_2 i$. Then $a + bj = a_1 + a_2 i + b_1 j + b_2 k$ using quaternion multiplication. The condition $|a|^2 + |b|^2 = 1$ becomes $|a_1|^2 + |a_2|^2 + |b_1|^2 + |b_2|^2 = 1$ hence $a + bj$ has quaternion norm 1. The map SU(2) $\rightarrow$ U is clearly a bijection. It remains to check that the map respects multiplication, so that it becomes a group isomorphism; this is a special case of Exercise 8.14.   QED

Theorem 8.5.2 (5) on the description of SO(3) can thus be reformulated as saying that there exists a two-to-one surjective group homomorphism SU(2) $\rightarrow$ SO(3) (compare also Exercise 8.3). The two groups are now matrix groups (over different fields), but the existence of the two-to-one map is by no means obvious from the matrix description: the most convincing way of going from complexes to reals is via quaternions.

8.7   The electron spin in quantum mechanics

This section relates the geometry of SO(3) to a fundamental attribute of elementary particles: their spin. All the mathematics needed is at hand already; however, there is no space in the present book to introduce all the necessary background from quantum mechanics. For more information and insight, see Feynman's classic [7], Chapters 1–3.

8.7.1

The story of the electron spin

The story begins in 1925. Two Dutch doctoral students George Uhlenbeck and Samuel Goudsmit, halfway through their Ph.D. program, noted that the electron inside the atom appeared to have, besides the three known 'quantum numbers' associated with the position of the electron, its angular momentum around the nucleus and its magnetic field, an extra degree of freedom. They postulated the existence of an extra 'quantum number', which they called the *electron spin*. This new quantum number seemed to behave in many ways like angular momentum, so they gave the interpretation that it corresponds to some kind of intrinsic rotational motion. However, the quantum number appeared to have just *two* possible values (+) and (−), and the rotation seemed not to have a definite axis; strange facts for a 'spinning' particle. Their advisor Paul Ehrenfest is said to have commented: 'You are both young enough to be able to afford a stupidity!' (he realised soon afterwards though that his students had in fact made an important discovery).

Unknown to Uhlenbeck and Goudsmit, the experimental verification of their discovery had been around for three years in the form of the Stern–Gerlach experiment. In 1922 the German scientists Otto Stern and Walther Gerlach built the device illustrated schematically in Figure 8.7a. The source emits a beam of silver atoms. The beam is directed between the poles of a magnet, which produces a magnetic field orthogonal to the direction of the path. As the atoms are electrically neutral, they are not expected to experience force; they should thus pass through the device without any change in their direction. However, a screen on the other side of the device

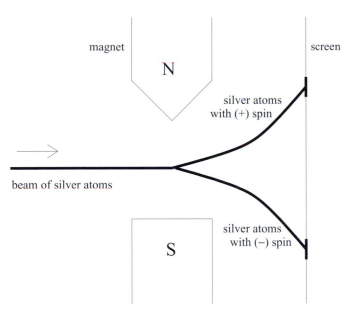

Figure 8.7a   The Stern–Gerlach experiment.

reveals that the atoms are in fact deflected by the magnetic field, and moreover that they follow one of two possible paths.

The experiment can only be understood in terms of the notion of spin. A silver atom has an electron on an outer shell, whose intrinsic spin interacts with the magnetic field. Atoms whose outer electron is in the (+) spin state follow a different path from those in the (−) spin state.

The mid-1920s was of course the time when quantum mechanics was invented. Soon after Uhlenbeck and Goudsmit's proposal, Pauli and Dirac incorporated electron spin into the quantum mechanical theory of the electron, also known as the Schrödinger equation. Since this is not a course about the electron, I do not need to worry unduly with the details.

8.7.2
Measuring
spin: the
Stern–
Gerlach
device

In the following, I assume a modified form of the Stern–Gerlach (SG) device, illustrated in Figure 8.7b. This is only a thought experiment[1], explained in detail in [7], pp. 5-1 and 5-2. An electron beam arrives from the left, and separates inside the device S into two beams according to its spin under the action of the left-hand 'magnet'. A combination of other 'magnets' forces the electrons back into their horizontal path; the outcoming beam still consists of a mixture of electrons in the two spin states.

Assume now that I block the path of one of the beams inside the device, as in the case of device S of Figure 8.7c. Then the electrons leaving the device S are all in a definite spin state (+). In this sense, I have now 'measured' the spin of this beam of

---

[1]   The experiment cannot be carried out as described here: the electron's wave function is too fuzzy because of quantum mechanical effects, and the separation into two rays is not apparent. The point about the silver atom featuring in the original Stern–Gerlach experiment is that it is electrically neutral, but has a relatively free electron on an outer shell; its motion between magnets is thus governed by the spin of the outer electron. In the text I stick to the thought experiment involving free electrons.

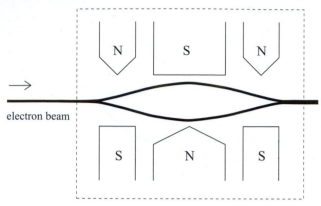

Figure 8.7b   The modified Stern–Gerlach device.

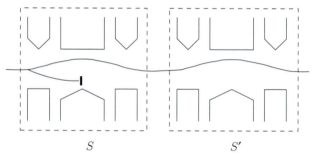

Figure 8.7c   Two identical SG devices.

electrons: I know precisely what state they are in. (Unfortunately, I have lost about half my electrons along the way, but that seems to be unavoidable in this kind of game. Compare with a large accountancy firm hired to count your money.) In particular, if I attach another SG device S' *in the same position* after the first as in Figure 8.7c, then I know the path of all the electrons inside the device; blocking the other path then makes no difference.

However, let us now put another SG device T *in a different spatial position* in the path of my uniform spin electron ray; see Figure 8.7d. The ray now *separates again*; the electrons choose two different paths in a specific ratio (which can be measured again by blocking one or other of the paths) depending on the position of the new SG device. Hence knowing that the electron is in spin state $(+)$ *in one direction* does not mean that it is in spin state $(+)$ in *all directions*. It registers as spin $(+)$ or $(-)$ in some different direction following, it seems, a fixed dress code.

8.7.3
The spin operator

As both experiment and speculation confirm, the electron spin takes two possible values $+1$ and -1, where I ignore unnecessary constants. In the framework of quantum mechanics, such a two-state system is modelled on a 2-dimensional complex vector space V with a definite Hermitian form on it, which I denote by bracket (,). Every electron in this simple model is described by its *wave function* $\psi \in V$, which we normalise to unit length $(\psi, \psi) = 1$.

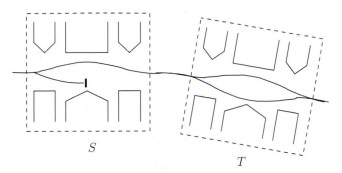

Figure 8.7d   Two different SG devices.

An SG device S in a fixed spatial position corresponds to a linear operator $\mathcal{O}_S: V \to V$. The possible spin states with respect to this spatial direction correspond to the different *eigenvalues* of this map. In the present case, the eigenvalues must therefore be ± 1. There are corresponding normalised eigenvectors ψ_S^+ and ψ_S^-:

$$\mathcal{O}_S(\psi_S^+) = \psi_S^+, \quad \mathcal{O}_S(\psi_S^-) = -\psi_S^-.$$

Quantum mechanics postulates that the operator $\mathcal{O}_S$ is *Hermitian* (Exercise 8.24). It follows that the eigenvectors are orthogonal $(\psi_S^+, \psi_S^-) = 0$. Thus $\{\psi_S^+, \psi_S^-\}$ is a Hermitian basis in the 2-dimensional vector space V.

The electron with wave function ψ_S^+ is in the $(+)$ spin state and that with wave function ψ_S^- is in the $(-)$ spin state. These electrons are in *eigenstates* of the spin operator $\mathcal{O}_S$. An arbitrary electron has a wave function $\psi \in V$ which is a linear combination of the basis vectors:

$$\psi = \alpha \psi_S^+ + \beta \psi_S^-.$$

Such a state is referred to as a *mixed state*.

An electron in a mixed state $\psi = \alpha \psi_S^+ + \beta \psi_S^-$ arriving at our SG device S passes along the $(+)$ or $(-)$ path in the device with probability $|\alpha|^2$ or $|\beta|^2$ respectively. These numbers are called *probability amplitudes*. Because both basis vectors $\psi_S^\pm$ and the vector ψ are normalised to unit length, $|\alpha|^2 + |\beta|^2 = 1$; thus these probabilities add to one.

Once we block the $(-)$ path, the outcoming electrons are all in the $(+)$ eigenstate: their wave function is the eigenvector $\psi_S^+ \in V$. This explains their behaviour in a next SG device S' in the same spatial position as S, pictured in Figure 8.7c. The operator corresponding to the device S' is $\mathcal{O}_{S'} = \mathcal{O}_S$, and the electrons are all in the $(+)$ eigenstate of this operator. So they choose the two paths with probability $|\alpha|^2 = 1$, respectively $|\beta|^2 = 0$; in other words, their path through S' is determined.

8.7.4
Rotate the
device

To perform our next thought experiment, imagine a beam of electrons leaving a device in one of the definite eigenstates, and arriving at another device in a different spatial position as in Figure 8.7d. The new SG device T corresponds to an operator $\mathcal{O}_T$ and hence to a new Hermitian basis $\{\psi_T^+, \psi_T^-\}$ of V consisting of eigenvectors of $\mathcal{O}_T$.

I wish to study an electron ray in one of the spin eigenstates $\psi_S^\pm$, when it passes through T. The experiment says that electrons will follow one of two possible paths in T, and I want the probability of its taking one or other of the paths. According to the rule spelled out in the last section, I should write the vector ψ_S^+ (and also ψ_S^-) in terms of the new basis $\{\psi_T^+, \psi_T^-\}$ to find the probability amplitudes. This is simply a change of basis, given by a 2×2 matrix $A_{S \to T}$, an element of GL(2, $\mathbb{C}$) (in fact U(2) as both bases are Hermitian). The task is to find $A_{S \to T}$ from S, T.

To proceed, I need to make precise the geometry of an SG device in 3-space. Note that an SG device in physical space $\mathbb{E}^3$ determines *two distinguished orthogonal directed lines*; namely, there is the distinguished direction of the electron beam, and the distinguished direction of the magnetic field orthogonal to it; see Figure 8.7a. I can think of these directed lines as two coordinate axes in a coordinate system, and there is a unique way of adding a third directed coordinate axis orthogonal to the first two to make a right-handed coordinate system in 3-space. The new system T determines in the same way a new right-handed coordinate system in $\mathbb{E}^3$. The transformation which gets me from S to T is a direct motion of $\mathbb{E}^3$, and thus a rotation $g \in$ SO(3). (Note that only directions matter in this discussion; the origin of the coordinate system is not important, and I ignore translations.)

According to the earlier discussion, I need a recipe associating an element of GL(2, $\mathbb{C}$) with a transformation $S \to T$, presumably in a continuous manner. In other words, I need a map

$$A: \ \mathrm{SO}(3) \to \mathrm{GL}(2, \mathbb{C}).$$

It can also be argued from basic principles of quantum mechanics that the map A should respect composition; after all, $S \to T$ followed by $T \to R$ should be the same as $S \to R$. Hence the map A should be a group homomorphism. This however presents a puzzle: there is no obvious way to map SO(3) to the group of linear maps on a 2-dimensional $\mathbb{C}$-vector space (apart from the map which takes every rotation to the identity matrix, which would contradict the experimentally observed fact that spin does depend on direction). In fact there is absolutely no such map at all.

8.7.5

The solution

Although the expressions for $\psi_T^\pm$ in terms of $\psi_S^\pm$ and the rotation taking S to T can be derived from first principles, I cannot improve on Feynman's beautiful and self-contained account (in pp. 6-1 to 6-14 of [7]), and I just state the result: namely, although there is no map $A:$ SO(3) $\to$ GL(2, $\mathbb{C}$), there is an obvious map

$$\widetilde{A}: \ \mathrm{SU}(2) \to \mathrm{GL}(2, \mathbb{C})$$

from the group SU(2) to GL(2, $\mathbb{C}$); a 2×2 unitary matrix is certainly invertible, so the inclusion map will do. On the other hand, SU(2) is not too different from SO(3); by Corollary 8.5.3, they are related by a two-to-one map. Thus $\widetilde{A}$ can be thought of as a two-valued function on SO(3).

Up to a knowledge of the explicit form of the map SU(2) $\to$ SO(3) that can easily be derived from the expressions in 8.5.2, this answers the original question of how to compute the ratio of electrons following the two paths of Figure 8.7d: $S \to T$ is given by an element of SO(3), and there are *two* possible changes of basis

$$\psi_T^+ = \alpha_+ \psi_S^+ + \beta_+ \psi_S^-$$
$$\psi_T^- = \alpha_- \psi_S^+ + \beta_- \psi_S^-$$

for matrixes

$$\begin{pmatrix} \alpha_+ & \alpha_- \\ \beta_+ & \beta_- \end{pmatrix} \in SU(2)$$

which differ from each other only in a change of sign; the eigenvectors are in any case determined only up to sign, and the physical meaning is only carried by the amplitudes $|\alpha_\pm|$ and $|\beta_\pm|$ which are independent of the choice of signs made.

One way to think of the process is to start with an SG device S and then start to turn it around a fixed axis. This determines a path in the group SO(3) starting from the identity. Starting from the identity matrix in SU(2), I can follow this path in SU(2), and see what happens to the transformation matrix. It turns out that after a full turn by 2π of my device, that is, after a loop in SO(3) returning to the identity, my path in SU(2) takes me to the *negative* of the identity matrix. Following the loop in SO(3) once again, I can continue my path in SU(2), and lo and behold! a turn of 4π returns me to the identity matrix in SU(2).

This thought experiment with paths reflects the topological fact that the fundamental group of SO(3) is $\mathbb{Z}/2$, and its universal cover is the map $S^3 \to$ SO(3) of 8.5–8.6 (see a first course in topology for the language). It is also responsible for the mysterious statement turning up frequently in physics texts, that 'rotation by 2π does not leave the wave function of the electron invariant, but multiplies it by (-1)'. As I am told, this can be directly demonstrated by experiment.

As a final comment, note that in this chapter I dealt with spin for a 'spin $\frac{1}{2}$' particle such as the electron, whose spin can take two values $(+)$ or $(-)$. There are also 'spin 1' particles such as the heavy particles Z, $W^\pm$ which are responsible for nuclear forces. Their spin can take the values $(+)$, 0 or $(-)$. Much of the discussion of this chapter applies to such three-state systems; compare [7], Chapter 5. Their spin can be measured by a three-way SG device. The vector space W representing spin states is now 3-dimensional over $\mathbb{C}$, and the transformation $S \to T$ between SG devices corresponds to a map $B \colon$ SO(3) $\to$ GL(3, $\mathbb{C}$). In this case, there is no great mystery: this map is, up to conjugation, the obvious inclusion map, where I think of a 3×3 real orthogonal matrix as a 3×3 complex invertible matrix (the 'vector representation'). For this reason spin 1 particles are often called 'vector particles'.

8.8 Preview of Lie groups

The topological groups GL(n, $\mathbb{R}$) and O(n) are examples of *Lie groups*, groups whose elements depend on a finite number of continuous parameters. Examples of Lie groups include the Euclidean group Eucl(n), the Lorentz group O$^+$(1, 2), the special linear group SL(n) (the group of invertible $n \times n$ matrixes with determinant 1), the spinor groups Spin(n), and groups defined using the complex numbers such as the group $GL(n, \mathbb{C})$ of invertible matrixes over $\mathbb{C}$. Here is a list of features of general Lie group theory that made an appearance in this chapter:

Dimension   The geometry of the group around any point can be described by d parameters, where the number d is independent of the point chosen, and is called the *dimension* of the group. Examples from Proposition 8.2 are

$$\dim O(n) = \binom{n}{2} \quad \text{and} \quad \dim \text{Eucl}(n) = \binom{n+1}{2}.$$

Components   A Lie group G has a number of connected components (finite or infinite), all of them geometrically the same (homeomorphic). The component containing the identity is a normal subgroup, and the other components are its cosets. See 8.4 for $O(n)$ and Exercise 8.5 for the group $O^+(1, 2)$.

Maximal compact subgroup   A connected Lie group G is homeomorphic to a product $H \times \mathbb{R}^N$ of a compact Lie group H and a space $\mathbb{R}^N$ in which all loops are contractible (compare 7.15). The examples of 8.3 are typical: *compactness* is achieved by imposing a *positive definite* orthogonal or Hermitian form.

The universal cover   A connected Lie group G has a cover $\widetilde{G} \to G$ by a simply connected Lie group $\widetilde{G}$ (possibly G itself). The typical examples are the exponential map $\mathbb{C} \to \mathbb{C}^*$ and the two-to-one spinor covers $S^3 \to SO(3)$ and $S^3 \times S^3 \to SO(4)$ discussed in 8.5.3.

Complexification and real forms   The group $GL(n, \mathbb{C})$ is the *complexification* of the group $GL(n, \mathbb{R})$: the latter is a matrix group, and I can simply take complex instead of real entries. Conversely, we say that $GL(n, \mathbb{R})$ is a *real form* of $GL(n, \mathbb{C})$. Along the same lines, the group $O(n, \mathbb{C})$ of $n \times n$ complex matrixes, which leave the standard quadratic (!) form $\sum_i x_i^2$ invariant, is a complexification of the group $O(n)$. However, $O(n)$ is not the only real form: over the complex numbers, there is no difference between the forms $\sum_i x_i^2$ and $-x_1^2 + \sum_{i>1} x_i^2$. Thus the Lorentz group $O(1, n-1)$ is also a real form of $O(n, \mathbb{C})$.

Linear representations   Just as finite groups, Lie groups are often studied via their linear (matrix) representations. In plain language, we associate to every group element $g \in G$ an $n \times n$ (complex) matrix A_g so that $A_h A_g = A_{hg}$. In fancier language, this is nothing but a group homomorphism $G \to GL(n, \mathbb{C})$; one familiar example is the map $\tilde{A}\colon SU(2) \to GL(2, \mathbb{C})$ from 8.7.5. I recommend Fulton and Harris [9] for further study.

Symmetry groups in physics   Lie groups commonly appear as symmetry groups of interesting physical systems. The mathematics of the group and the physics of the system are often related in beautiful and nontrivial ways. The interaction occurs on two levels: 'classical' (meaning Newtonian dynamics and Maxwell electromagnetic theory) and 'modern' (meaning relativity theory or quantum mechanics, possibly both). The story of the electron in 8.7.5 is the starting point of the 'quantum' level of this interaction; for more discussion, turn to 9.3 and Sternberg [23].

Exercises

8.1 How much bigger is the affine group Aff(n) than the Euclidean group Eucl(n)? [Hint: compare GL(n) and O(n) in 8.3.]

8.2 (a) Show that rotations, translations, reflections and glides of $\mathbb{E}^2$ (Theorem 1.14) depend respectively on 3, 2, 2 and 3 parameters.
   (b) Count parameters for each of the types of motion of Theorem 1.15. (Answers: (1) translation 3; (2) rotation 5; (3) twist 6; (4) reflection 3; (5) glide 5; (6) rotary reflection 6. For example, a rotation is specified by a line of 3-space, which depends on 4 parameters, plus an angle.)

8.3 Count the number of *real* parameters for the groups SO(3) and SU(2); verify that they depend on the same number of parameters, as you would expect from the two-to-one cover discussed in 8.6. [Hint: use Proposition 8.2, respectively the results of 8.6.]

8.4 Determine the connected components of GL($n, \mathbb{R}$) using Theorem 8.3 and Proposition 8.4.

8.5 Let

$$O(1, 2) = \left\{ A \in GL(3, \mathbb{R}) \,\middle|\, {}^t A J A = J \right\}$$

be the group of all Lorentz matrixes, which contains the Lorentz group $O^+(1, 2)$ introduced in 8.1, Example 5. Show that this group has *four* connected components, distinguished by whether a matrix preserves the cone $q_L(\mathbf{v}) < 0$ or maps it to $q_L(\mathbf{v}) > 0$ (that is, whether it is in $O^+(1, 2)$), and det $A = \pm 1$. [Hint: imitate the proof of Proposition 8.4, using the Lorentz normal form statement of Exercise B.3. Distinguish carefully between four types of possible diagonal matrixes arising as end products.]

8.6 Let $A \in GL(n, \mathbb{R})$ be a matrix with columns $\mathbf{f}_i$. Following the proof of Theorem B.3 (1) carefully, show that it is possible to construct an orthonormal basis $\{\mathbf{e}_i\}$ of $\mathbb{R}^n$, so that in each step

$$\mathbf{e}_i = c_{i1}\mathbf{f}_1 + \cdots + c_{ii}\mathbf{f}_i$$

with $c_{ii} > 0$. Let $C = (c_{ij})$ and B the matrix with columns $\mathbf{e}_i$; check that $A = BC$ and that $B \in O(n), C \in T_+(n)$ (compare 8.3). Check also that the entries of B and C depend continuously on those of A.

8.7 Write the following matrixes in the form BC of Theorem 8.3 with $B \in O(n)$ and $C \in T_+(n)$:

$$\begin{pmatrix} 1 & 1+\sqrt{3} \\ \sqrt{3} & -1+\sqrt{3} \end{pmatrix}, \quad \begin{pmatrix} 1 & 3 \\ 1 & 4 \end{pmatrix}, \quad \begin{pmatrix} 1 & 0 & 3 \\ 2 & -1 & 4 \\ 2 & 1 & 2 \end{pmatrix}.$$

Exercises on quaternions.

8.8   Show that 4 complex matrixes

$$1 = \begin{pmatrix} 1 & 0 \\ 0 & 1 \end{pmatrix}, \quad I = \begin{pmatrix} i & 0 \\ 0 & -i \end{pmatrix}, \quad J = \begin{pmatrix} 0 & 1 \\ -1 & 0 \end{pmatrix}, \quad K = \begin{pmatrix} 0 & i \\ i & 0 \end{pmatrix}$$

multiply together by the same rules as the 4 basic quaternions $1, i, j, k$. Since matrix multiplication is associative, use this to give a better proof of Proposition 8.5.1 (1).

8.9   Complete the proof by brute force of $(pq)^* = q^* p^*$ for quaternion conjugation (Proposition 8.5.1 (2)). Give a better proof along the lines of the previous exercise.

8.10   Study the group $G_8 = \{\pm 1, \pm i, \pm j, \pm k\}$ of unit quaternions. Write out the group multiplication table, and find a convincing reason (or failing that, any reason) why G_8 is not isomorphic to the dihedral group D_8 appearing in Exercise 6.5.

8.11   If $p = ai + bj + ck$ and $q = di + ej + fk$ are two pure imaginary quaternions, calculate $pq + qp$ directly using the definition of quaternion multiplication.

8.12   Prove that a pure imaginary quaternion p satisfies $p^2 = -|p|^2$. Also if p, q are pure imaginary then $pq + qp = 0$ if and only if they are orthogonal with respect to the quadratic form $a^2 + b^2 + c^2 + d^2$. [Hint: orthogonal with respect to a quadratic form Q is expressed in terms of the associated bilinear form $\varphi(p, q) = Q(p + q) - Q(p) - Q(q)$; apply this with $Q(q) = qq^* = -q^2$.]

       Deduce that 3 vectors $I, J, K \in \mathbb{H}$ have the same multiplication table as the quaternion basis i, j, k if and only if they are an oriented orthonormal frame of $\mathbb{R}^3$. Prove Proposition 8.5.1 (5).

8.13   Show how to express $\mathbb{C}$ in terms of 2×2 matrixes over $\mathbb{R}$ of the form $\begin{pmatrix} a & b \\ -b & a \end{pmatrix}$.

8.14   Show that the algebra of 2×2 matrixes over $\mathbb{C}$ of the form $\begin{pmatrix} a & b \\ -\bar{b} & \bar{a} \end{pmatrix}$ is an algebra isomorphic to the quaternions $\mathbb{H}$. [Hint: consider the basis given in Exercise 8.8 and compare also 8.6.]

8.15   Consider left multiplication by $M = \begin{pmatrix} a+ib & c+id \\ -c+id & a-ib \end{pmatrix}$ acting on $\mathbb{C}^2$. Write out the action of M on $\mathbb{C}^2 = \mathbb{R}^4$ in terms of the $\mathbb{R}$-basis $(1, 0), (i, 0), (0, 1), (0, i)$ of $\mathbb{C}^2$. Prove that the determinant of the map on $\mathbb{R}^4$ is $(a^2 + b^2 + c^2 + d^2)^2$. Use this to give another proof that a_q is direct in Theorem 8.5.2 (1).

8.16   Prove that 2×2 matrixes over $\mathbb{R}$ of the form $\begin{pmatrix} a & b \\ b & a \end{pmatrix}$ form an algebra B, and study its properties. Why is it not very interesting? [Hint: show that B is closed under addition and multiplication of matrixes. Find a basis over $\mathbb{R}$, and write out the multiplication table.]

8.17   By analogy with the previous question, investigate the algebra of 2×2 matrixes over $\mathbb{C}$ of the form $\begin{pmatrix} a & b \\ -b & a \end{pmatrix}$.

8.18   Use the argument of Theorem 8.5.2 to find a unit quaternion q so that the rotation $r_q : x \mapsto qxq^*$ is $(x, y, z) \mapsto (y, -x, z)$.

8.19   Find a unit quaternion q so that the rotation $r_q : x \mapsto qxq^*$ is $x \mapsto y \mapsto z \mapsto x$. [Hint: the effort intensive method is to use brute force. The thinking person's method is to represent $x \mapsto y \mapsto z$ as a rotation through angle θ about directed axis L, then use Theorem 8.5.2.]

8.20   By analogy with 1.11.1, solve the relations (1) of 8.6 to get $d = \overline{a}$, $c = -\overline{b}$. [Hint: for example, do second line $\times$ d − third line $\times$ c, then substitute $ad - bc = 1$ on the right-hand side.]

8.21   (Harder) Using the results of the two preceding exercises, show how to find a subgroup BO_{48} of the unit quaternions which has a surjective two-to-one map to the group of rotations of the cube in SO(3).

8.22   (Harder) Complete the proof of Theorem 8.5.2 (2).

   (a) Prove that $\varphi(p, q) = \mathrm{id}_{\mathbb{H}}$ if and only if $(p, q) = (1, 1)$ or $(p, q) = (-1, -1)$. [Hint: $p1q^* = 1$ if and only if $p = q$, and $pip^* = i$ if and only if $pi = ip$ if and only if $p = a + bi$, etc.] Deduce that φ induces an injective map $(S^3 \times S^3)/\pm 1 \rightarrow$ SO(4).

   (b) Prove that φ is surjective. [Hint: find a suitable $\varphi(p, q)$ to send 1 to a given unit vector $r \in \mathbb{H}$. Now compose with r^* to assume that $1 \mapsto 1$, and apply Theorem 8.5.2 (4).]

8.23   (Harder) Consider the algebra $\mathbb{O}$ of 2×2 matrixes over the quaternions $\mathbb{H}$ of the form $\left(\begin{smallmatrix} a & b \\ -b^* & a^* \end{smallmatrix} \right)$ where a^* is the quaternion conjugate of a as in 8.5.1.

   (a) Show that $\mathbb{O}$ is an 8-dimensional division algebra (algebra with two-sided multiplicative inverses for nonzero elements) over $\mathbb{R}$. Find an explicit basis for $\mathbb{O}$ and write out some of the multiplication table.

   (b) Show that multiplication in $\mathbb{O}$ is not associative, but it satisfies the identity

$$x(xy) = (xx)y \text{ for } x, y \in \mathbb{O}.$$

   (c) Contemplate on the possibility of doing projective geometry over the division algebra $\mathbb{O}$ (compare the end of 5.12).

   $\mathbb{O}$ is the algebra of *Cayley numbers* or *octonions*. For much more on this, see Conway and Smith [4].

   [Hint: you get a division algebra by introducing an octonion conjugate $\widehat{a}$ such that $a\widehat{a} = |a|^2$ is positive definite, as in 8.5.1. It is easy to find examples of nonassociative octonion multiplication; to prove the weaker identity, one possibility is to use your basis for $\mathbb{O}$ over $\mathbb{R}$ in a brute-force proof similar to that of Proposition 8.5.1 (1) given in the text. To do projective geometry, you have to start by thinking about the relation $\mathbf{x} \sim \lambda\mathbf{x}$ used to define projective space. Do not be surprised if you run into difficulty.]

Hermitian matrixes.

8.24   An $n \times n$ complex matrix A is called *Hermitian*, if $^h A = A$. (See 8.6 for the Hermitian conjugate $^h A$.) Show that

   (a) every eigenvalue of a Hermitian matrix is real;

   (b) eigenvectors for different eigenvalues are orthogonal with respect to the Hermitian form on $\mathbb{C}^n$ (compare Step 3 in the proof of Theorem 1.11!).

9 Concluding remarks

This final chapter is quite different from the earlier ones in style and intention: I let my hair down with a number of informal fairy stories on different topics, tying together loose strands in the historical and mathematical argument of the book, and opening up some new directions. In particular, I give a 'popular science' discussion of some of the surprising and amazingly fertile links between the geometry, topology and Lie group theory discussed in this book and different aspects of twentieth century physics.

There are many other topics closely related to the main text, both frivolous and serious, that I would have liked to write about. But life is short, and I confine myself to a brief list of a few directions and developments. Several of these topics can form the basis for undergraduate essays or projects.

- The classification of locally Euclidean geometries in the style of Nikulin and Shafarevich [18].
- Spherical trig and geometry in the history of navigation. Modern developments: GPS (global positioning system) devices.
- Spherical geometry and cartography (map making): Mercator's and other projections, as discussed for example in [6].
- Plane and spherical geometry and plate tectonics, following for example [8], Chapter 2. Why South America and West Africa fit together like pieces of a spherical jigsaw puzzle; Euler's theorem and the classification of fault types.
- SO(3) and Euler angles, mechanics in moving frames, Coriolis forces.
- Symmetry groups in geometry. This is a vast subject, relating regular polyhedra and polytopes, crystallography [5, 18], the geometric patterns of the Alhambra and other Islamic art, Escher's art and Penrose tilings.

- Subgroups of the symmetric group in puzzles and toys. Examples include the perfect shuffle groups and moves of the Rubik cube, as in [17] Chapter 19.
- Axiomatic projective geometry, leading to von Neumann's foundations of quantum theory, $\mathbb{C}^*$ algebras and 'noncommutative geometry'.
- Geometry and dynamics: Newton's equations, planetary motion and conics.
- Differential geometry of curves and surfaces. The Frénet frame, intrinsic curvature and the Gauss–Bonnet formula.

I leave you to explore details of these fascinating topics, as well as those sketched below, in or out of the confines of a degree course and its attendant examinations.

9.1 On the history of geometry

9.1.1 Greek geometry and rigour

Geometry has a very special place in the history and culture of western mathematics. Coming at the dawn of western civilisation (350 ± 200 BC), Greek philosophy and geometry, passed on to us by the more advanced culture of the Islamic world at the time of the Renaissance, has played a central role in the development of western culture, not merely for its content, but for its idea of rigour. The Greeks were not the first to attempt to describe the world around them by 'geometry': that credit goes to the ancient Mesopotamians (from 2500 BC), followed by the Egyptians (from 2000 BC). However, before the Greeks, geometry largely consisted of a bag of tricks for calculation that worked in practice most of the time. In contrast, Greek mathematicians elaborated the notion of logical argument. By this I do not mean the elementary and often hairsplitting logic of a 'Foundations' or 'Set theory' or 'Abstract algebra' course, but the idea that understanding steps at different stages in an argument from the ground up is at least as important as somehow getting an approximately correct answer. This is one of the fundamental items of intellectual equipment that set western mathematics and science apart from (and in the course of time well above) that of India and China.

Building on sources largely unknown to us, the geometer Euclid, probably working in Alexandria in the fourth century BC, summarised the mathematical knowledge of the time in his 13 volume *Elements*. Book I deals with the basic definitions of geometry. Euclid introduces notions such as *point*, *line*, *plane*, *distance*, *angle* and *meets*, whose meaning is supposed to be self-evident, and enunciates certain *postulates* (in modern language, axioms) concerning these notions. Lengths and angles are to be thought of as geometric quantities in their own right, not related to any algebraic or numeric representation. For example, one of the postulates states that two line segments are equal if they are congruent, which makes perfect sense without having to consider the length of a line as a number.

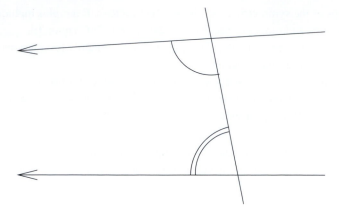

Figure 9.1a   The parallel postulate. To meet or not to meet?

9.1.2
The parallel
postulate

Most of Euclid's postulates were for a long time beyond doubt, but the last one stood out from the beginning as far less obvious:

> If a line falls on two lines, with interior angles on one side adding to less than two right angles, the two lines, if extended indefinitely, meet on the side on which the angles add to less than two right angles.

This is nonobvious. Behold Figure 9.1a! Euclid's 'extended indefinitely' makes it clear that the statement involves arguing on objects that are arbitrarily distant, so that it is in principle not verifiable. Through the ages, many alternative axioms were formulated, which can be proved to be equivalent to Euclid's on the basis of the other axioms, such as:

> given a line L in the plane, and a point P not on L, there exists one and only one line through P not meeting L

(compare Figure 9.1b and Figure 3.13). Or

> the sum of the angles of a triangle is equal to two right angles

(see Figure 1.16b and Theorem 3.14).

After arguably the longest dispute in intellectual history, it was discovered between about 1810 and 1830 by Bolyai, Gauss, Lobachevsky and Schweikart (independently, alphabetical order) that the parallel postulate cannot be a consequence of Euclid's other axioms: axiomatic geometries exist which are in many ways similar to Euclidean plane geometry, sharing its aesthetic appeal and simplicity, but which do not satisfy the parallel postulate. As János Bolyai wrote to his father,

> ollyan felséges dolgokat hoztam ki, hogy magam elbámultam, s örökös kár volna elveszni; ha meglátja Édes Apám megesméri; most többet nem szólhatok, tsak annyit: hogy semmiből egy ujj más világot teremtettem; mind az, valamint eddig küldöttem, tsak kártyaház a toronyhoz képest...

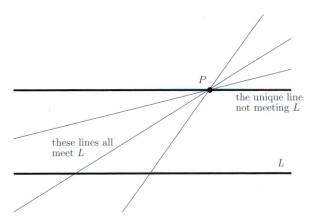

the unique line
not meeting L

P

these lines all
meet L

L

Figure 9.1b The parallel postulate in the Euclidean plane.

Or, translated from the nineteenth century Hungarian:

> I deduced things so marvellous that I was enchanted myself, and it would be an eternal loss to let them pass; Dear Father, once you see them, you will recognise their greatness yourself; now I cannot tell you more, only this: out of the void I created a new, a different world; all that I sent you before is like a house of cards to a tower. . .

The discovery of non-Euclidean hyperbolic geometry was indeed a landmark in modern scientific thinking, as revolutionary and as far reaching in its implications as the Copernican model of the solar system or Darwin's theory of evolution. For an account of the very interesting history, see Greenberg [11] and Bonola [3]. The early models of hyperbolic geometry were abstract; simple coordinate models, such as that used in Chapter 3 of this course, were developed later in the second half of the nineteenth and the early twentieth centuries. As I said, the coordinate model of hyperbolic geometry constructed in Chapter 3 satisfies all of Euclid's postulates except for the parallel postulate; the parallel postulate is therefore certainly not a logical consequence of the others. Hyperbolic geometry soon found many applications in different areas of mathematics and science; in particular, the notion of curvature in differential geometry and of curved space plays a foundational role in Einstein's general relativity (1916).

Spherical geometry seems to have been excluded from consideration in descriptive or axiomatic geometry from the time of Euclid for two reasons.

(a) More obviously, any two lines meet in two points (a pair of antipodal points); this is not a very serious defect, because you can pass to the geometry of $S^2/\{\pm 1\} = \mathbb{P}^2_{\mathbb{R}}$, in which every pair of lines meets in just one point.

(b) Its lines do not satisfy the *order* condition implicit in Euclid: given three points P, Q, R on a spherical line (great circle), it is impossible to say which of the three is 'between' the other two. Equivalently, a point P of a spherical line (great circle) does not divide it into disconnected sets. That is, given a line L and a point P not on it, *every* line M through P meets L *both* over there to the left *and* over there to the right

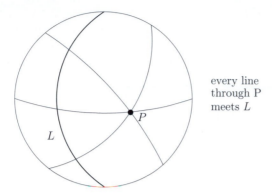

Figure 9.1c   The 'parallel postulate' in spherical geometry.

(see Figure 9.1c). In spherical geometry these are antipodal points; in the geometry of $S^2/\{\pm 1\} = \mathbb{P}^2_{\mathbb{R}}$, the same point.

Euclid's postulates did not discuss the separation properties of points on a line: it was supposed to be understood what it meant for A to be between P and Q on the line segment PQ. (Compare the discussion in 7.3.3; separation is a topological statement about the geometry.) Thus it is not surprising that spherical geometry was overlooked; however, this is a fair indication that Euclid's claim to rigour in a modern sense was never really watertight.

Nevertheless, spherical geometry has been around in an 'applied' form for centuries. Spherical trigonometry was studied in amazing detail by the great medieval Islamic geometers in the context of *qibla* (the sacred direction to Mecca, see for example [16], and again from the time of Newton, to aid British ships engaged in piracy or the slave trade to navigate around the oceans of the world and return to the other origin at Greenwich. Because of winds and currents though, the lines of spherical geometry, great circles, are not always the fastest way to travel. These days, great circles are the routes taken for preference by airlines, except when no-fly zones intervene.

**9.1.3
Coordinates
versus
axioms**

Descartes' invention of coordinate geometry is another key ingredient in modern science. It is scarcely an accident that calculus was discovered by Leibnitz and Newton (independently, alphabetical order) in the fifty years following the dissemination of Descartes' ideas. Interactions between the axiomatic and the coordinate-based points of view go in both ways: coordinate geometry gives models of axiomatic geometries, and conversely, axiomatic geometries allow the introduction of number systems and coordinates. There are several excellent books giving systematic treatments of these very interesting issues; I warmly recommend Hilbert's classic [13].

As in art or music or politics, attitudes and fashions in mathematics vary quite sharply from one generation to the next. In the second half of the nineteenth century, up to the time of Hilbert and Poincaré, geometry was without doubt at the centre of mathematics and of large areas of theoretical physics. This position was overturned with the rise of abstract algebra, topology and set theoretic foundations of mathematics around the 1920s. The blame for this lies in part with the geometers themselves, who developed a sloppy attitude to correct statements and proofs of theorems. One example is the type of argument that involved a 'sufficiently general position', which might in favourable cases have a precise meaning within an epsilon neighbourhood of the author. In England, there was a brilliant school of geometers between the wars in Cambridge, which seems to have been broken up when the participants were drafted into code breaking or aeronautics during the second world war. When the senior author was an undergraduate at Cambridge (late 1960s), geometry in the sense of this course was universally considered a terribly dull fuddy-duddy subject. The position has been entirely turned around in the last 30 years, and at present geometry in its various manifestations again claims centre stage in mathematics and theoretical physics.

9.2 Group theory

**9.2.1
Abstract
groups
versus
transforma-
tion
groups**

According to the abstract definition (which is comparatively recent), an abstract group is a set with a composition law satisfying a couple of well known axioms. However, from the beginnings of the subject in the nineteenth century, the groups studied were always thought of as symmetry groups, that is, as transformation groups preserving some structure or other. For example, Ruffini, Abel and Galois considered permutations of the roots of a polynomial equations, and the subgroup of permutations that preserve the rules of arithmetic. From the mid-nineteenth century, many other groups arose as geometric symmetries: finite groups such as the symmetries of the regular polyhedra, infinite but discrete groups in the study of crystallography, that contain translations by a lattice as a subgroup, and Lie groups such as the Euclidean group. The idea that a group can be treated as an abstract composition law *without reference to the nature of the operators that make it up* was first introduced by Cayley in 1854, but its significance was not recognised until much later.

Let G be a group and Σ a set; I say that Σ is a G-set or that G *acts on* Σ, if a group homomorphism

$$\varphi : G \to \text{Trans } \Sigma$$

is given from G to the group of transformations of Σ (see 6.1). That is, each $g \in G$ corresponds to a transformation (bijective map) $\varphi_g : \Sigma \to \Sigma$, in such a way that the abstract composition law in G corresponds to composition of transformations of Σ. In other words, G is trying to fulfil its destiny as a transformation group of Σ, as discussed in Chapter 6. One usually writes simply $\varphi_g(x) = gx$ or $g(x)$ for the action of $g \in G$ on $x \in \Sigma$.

The requirement that the map φ is a homomorphism is written $(gh)x = g(hx)$. This looks like an associative law, but it just means that the abstract product in G corresponds to composition of maps $\Sigma \to \Sigma$; compare the discussion in 2.4. Evaluating $g \in G$ on $x \in \Sigma$ provides a map $\Phi : G \times \Sigma \to \Sigma$ given by $\Phi(g, x) = \varphi_g(x)$; I leave it to you to express the condition $(gh)x = g(hx)$ in these terms.

9.2.2
Homo-
geneous and
principal
homo-
geneous
spaces

Definition   Let Σ be a G-set. I say that G acts *transitively* on Σ if the action takes any point of Σ to any other. In this case Σ is a *homogeneous space* under G.

This idea has already appeared many times: the geometries in the earlier chapters of the book were homogeneous under appropriate groups. For example, the Euclidean group acts transitively on $\mathbb{E}^n$: any point of $\mathbb{E}^n$ goes to the origin under a suitable Euclidean motion. The affine group $\mathrm{Aff}(n)$ acts transitively on *pairs* of distinct points of $\mathbb{A}^n$; as discussed at several points of the book, this is closely related to the fact that affine geometry does not have an invariant distance function.

If Σ is a G-set and $x \in \Sigma$, the *stabiliser subgroup* of x is the set of elements of G that fix x, that is

$$\mathrm{Stab}_G(x) = \big\{ h \in G \mid h(x) = x \big\}.$$

For example, the stabiliser subgroup of the origin $0 \in \mathbb{E}^n$ in $\mathrm{Eucl}(n)$ is the group $\mathrm{O}(n)$ of orthogonal matrixes.

If G acts transitively on Σ, the map $e_x \colon G \to \Sigma$ defined by $g \mapsto gx$ is surjective. Moreover elements $g_1, g_2 \in G$ map to the same point of Σ if and only if $g_2 = g_1 h$ for some $h \in \mathrm{Stab}_G(x)$; thus e_x induces a bijection $G/\mathrm{Stab}_G(x) \xrightarrow{\cong} \Sigma$. (Here G/H stands for the quotient of G by the equivalence relation $g \sim gh$ for $h \in H$, or the set of left cosets of H.)

Definition   A homogeneous space Σ under G is a *principal homogeneous space* under G or a *G-torsor* if the stabiliser $\mathrm{Stab}_G(x)$ is trivial for every $x \in \Sigma$. Since the stabilisers of x and gx are conjugate (by the same argument as in Exercise 6.7), it is enough to verify that $\mathrm{Stab}_G(x)$ is trivial for a single $x \in \Sigma$.

For example, affine space $\mathbb{A}^n$ is a homogeneous space under $\mathrm{Aff}(n)$, but is a torsor under the translation subgroup $\mathbb{R}^n \subset \mathrm{Aff}(n)$.

According to the previous discussion, if Σ is a G-torsor, then $e_x \colon G \to \Sigma$ is a bijection from G to Σ, and I could use this to identify G and Σ. However, different elements of Σ give different bijections: the set Σ has no distinguished identity element.

Example   Let Σ consist of the vertexes of a regular n-gon in the plane $\mathbb{E}^2$, $G \subset \mathrm{Eucl}(2)$ the group of symmetries of Σ (the dihedral group D_{2n}, see Exercise 6.5), and let H be the cyclic subgroup of G of order n consisting of rotations. (Draw a picture!) Then the geometric action of G on Σ is transitive, since the polygon is regular. Thus Σ is a homogeneous space under G. The stabiliser $\mathrm{Stab}_G(P)$ of a vertex $P \in \Sigma$ is of order two, consisting of the identity and the reflection in the axis through P. The subgroup H acts transitively and without stabilisers (since it does not contain reflections). Thus Σ is an H-torsor: there are as many vertexes as rotations, but no vertex is distinguished over the others.

9.2.3
The
Erlangen
program
revisited

Recall Klein's Erlangen program of Section 6.3: the slogan is that geometry is the study of properties invariant under a transformation group G. The introduction to Chapter 1 discussed the basic geometric and philosophical principles: space should be

(1) homogeneous (the same viewed from every point), and
(2) isotropic (the same in every direction).

In terms of the group of transformations, (1) says that the group G acts transitively on points of space, whereas (2) says that it also acts transitively on coordinate frames based at every point. Helmholtz' axiom of free mobility requires slightly more: it also says that, given two points of the space and sets of coordinate frames based at these points, there is a *unique* element of G mapping one to another. In other words, the set of all coordinate frames at all points is a G-torsor (principal homogeneous space under G). Thus

- Euclidean space $\mathbb{E}^n$ is a homogeneous space under the Euclidean group $\mathrm{Eucl}(n)$. The stabiliser of a point $P \in \mathbb{E}^n$ is isomorphic to the group $\mathrm{O}(n)$, the group of rotations and reflections fixing P. By Theorem 1.12, the set of Euclidean frames forms a torsor under $\mathrm{Eucl}(n)$.
- The sphere S^n is a homogeneous space under the group $\mathrm{O}(n+1)$ of spherical motions (Theorem 3.4 for $n = 2$; the general case is identical). For $P \in S^n$, the stabiliser group is isomorphic to the group $\mathrm{O}(n)$. (It is the group of orthogonal matrixes in the $\mathbb{R}^n$ that is the orthogonal complement of OP.)
- Hyperbolic space $\mathcal{H}^n$ is homogeneous under the Lorentz group $\mathrm{O}^+(n, 1)$. The stabiliser of a point P is again isomorphic to the group $\mathrm{O}(n)$.
- Projective space $\mathbb{P}^n$ is homogeneous under the projective linear group $\mathrm{PGL}(n+1)$. The stabiliser of a point $P \in \mathbb{P}^n$ is $\mathrm{PGL}(n)$. By Theorem 5.5, the set of projective frames of reference forms a $\mathrm{PGL}(n+1)$-torsor.

9.2.4
Affine space
as a torsor

The notion of torsor formalises the ad hoc definition of affine space I gave in Chapter 4. Let V be a vector space; an affine space $\mathbb{A}(V)$ is just a torsor under V. In other words, $\mathbb{A}(V)$ is a set with an action of V ('by translation'), and this action is simply transitive: for $P, Q \in V$ there is a unique vector $\mathbf{x} \in V$ such that $Q = P + \mathbf{x}$.

Looking back to 6.5.3, I can say all this slightly differently: the transformation groups in Euclidean and affine geometry are semidirect products. For example, the Euclidean group

$$\mathrm{Eucl}(n) = \mathrm{O}(n) \ltimes \mathbb{R}^n$$

is the semidirect product of the normal subgroup of translations and the group of rotations. From the analysis of 6.5.3, it follows that the subgroup $\mathrm{O}(n)$ is not normal. The conjugation construction (see 6.4) allows me to *define* Euclidean space to be the space of all conjugates of a fixed copy of $\mathrm{O}(n) \subset \mathrm{Eucl}(n)$, and notions of Euclidean geometry to be all notions that can be defined on this space invariantly under the group $\mathrm{Eucl}(n)$. This is of course the Erlangen program repeated once again.

I can say the same words starting from the group of affine transformations $\mathrm{Aff}(V)$ (see 4.5). This contains copies of $\mathrm{GL}(V)$, the group of invertible linear maps of V, as affine transformations fixing a point, and these subgroups are once again nonnormal.

From the group theory it follows then that the group of translations V acts transitively with trivial stabiliser on $\mathbb{A}(V)$; thus $\mathbb{A}(V)$ is a V-torsor (a principal homogeneous space under the group of translations). In other words, we have an action $\varphi_{\mathbf{v}} : P \mapsto P + \mathbf{v}$ of the additive group of V defined on points of affine space. For $P \in \mathbb{A}(V)$, we get a bijection $e_P : V \to \mathbb{A}(V)$ mapping $\mathbf{v} \in V$ to $P + \mathbf{v}$; two such identifications differ by an element of V acting by translation. The bijections e_P are different coordinate systems on affine space, differing by a translation; in the coordinate system e_P, the point P plays the role of origin. We also see that two points $P, Q \in \mathbb{A}(V)$ determine a vector $e_P(Q) = \overrightarrow{PQ} \in V$ (cf. Figure 4.2).

The point here is that for the cases I am interested in, I can recover the geometry from the group or the group from the geometry. For example, if the Euclidean group $\mathrm{Eucl}(n)$ and its subgroup $\mathrm{O}(n)$ are given, $\mathbb{E}^n$ is the homogeneous space $\mathrm{Eucl}(n)/\mathrm{O}(n)$, where $\mathrm{O}(n) = \mathrm{Stab}(x)$; alternatively, $\mathbb{E}^n$ is the set of subgroups conjugate to $\mathrm{O}(n)$.

9.3   Geometry in physics

Some of the most substantial applications of geometric ideas come from physics. Recall the grandiose aim expressed in my first sentence:

Geometry attempts to describe and understand space around us and all that is in it.

You may well object that most of the work so far has gone into describing the space, so it is about time I told you something about what is in it. The discussion is necessarily somewhat sketchy and in places wildly over-simplified; at the end I give references to the literature for further study.

9.3.1
The Galilean group and Newtonian dynamics

The dynamics of Galileo and Newton takes Euclidean three space $\mathbb{E}^3$ as the fundamental model of physical space, and time t as a universal parameter with a preferred directionality. Thus spacetime is modelled by $\mathbb{E}^3 \times \mathbb{R}$, with coordinates $(\mathbf{x}, t)$. Spatial lengths are measured with respect to the Euclidean metric of 1.1, and involve only the $\mathbf{x}$-coordinate; events also have a time separation $t_2 - t_1$ (no absolute value is taken here). Valid coordinate systems describing Newtonian dynamics are based on inertial frames in uniform relative motion with respect to each other, in which spatial lengths and time differences are unchanged. Transformations to a different coordinate system are therefore given by maps

$$(\mathbf{x}, t) \mapsto (A\mathbf{x} + \mathbf{g}t + \mathbf{b}, t + s),$$

where $A \in \mathrm{O}(3)$ is a 3×3 orthogonal matrix, $\mathbf{g}$ and $\mathbf{b}$ are 3×1 column vectors, and $s \in \mathbb{R}$ is a scalar. Such transformations collectively form the *Galilean group* $\mathrm{Gal}(3, 1)$ of classical $(3 + 1)$-dimensional spacetime $\mathbb{E}^3 \times \mathbb{R}$. A simple parameter count shows that the Galilean group depends on $3 + 3 + 3 + 1 = 10$ parameters. You recognise $\mathrm{Eucl}(3)$ as a subgroup of $\mathrm{Gal}(3, 1)$ consisting of time-independent transformations

Table 9.3 *Symmetries and conservation laws*

Symmetry	Conserved quantity	Name
spatial translation $(\mathbf{x}, t) \mapsto (\mathbf{x} + \mathbf{b}, t)$	$\sum_i m_i \dfrac{d\mathbf{x}_i}{dt}$	momentum
spatial rotation $(\mathbf{x}, t) \mapsto (A\mathbf{x}, t)$	$\sum_i m_i \mathbf{x}_i \times \dfrac{d\mathbf{x}_i}{dt}$	angular momentum
Galilean boost $(\mathbf{x}, t) \mapsto (\mathbf{x} + \mathbf{g}t, t)$	$-\mathbf{p}t + \sum_i m_i \mathbf{x}_i$	centre of mass (where $\mathbf{p}$ is the total momentum)
time translation $(\mathbf{x}, t) \mapsto (\mathbf{x}, t + s)$	$\sum_i \dfrac{1}{2} m_i \left\lvert \dfrac{d\mathbf{x}_i}{dt} \right\rvert^2$	energy

$$(\mathbf{x}, t) \mapsto (A\mathbf{x} + \mathbf{b}, t),$$

with $\mathbf{g} = 0$ and $s = 0$. Transformations with nonzero $\mathbf{g}$ correspond to a change to a new reference frame in uniform movement of speed $\mathbf{g}$ with respect to the old one; such group elements are usually called *Galilean boosts*. Elements of Gal(3, 1) with $s \neq 0$ correspond to moving the origin of time; Newtonian physics has no fixed Creation or Big Bang. It is however not possible to stretch or reverse time, however much you might wish it during an exam.

The shape of the Galilean group determines Newton's equation of motion, in the form familiar to you from a first mechanics course. For a single particle with mass m and position vector $\mathbf{x}(t)$ at time t, with no external forces acting, the equation simply says

$$m \frac{d^2 \mathbf{x}(t)}{dt^2} = 0.$$

Note that this equation is indeed invariant under the Galilean group.

Emmy Noether's principle of conserved quantities says that for a physical system with a symmetry group, there are as many conserved quantities (constants of the system unchanged as a function of time) as parameters for the group. As noted above, the Galilean group depends on 10 parameters, so we are looking for 10 conserved quantities. For a system with n particles having masses m_i and position vectors $\mathbf{x}_i(t)$, Table 9.3 describes the conserved quantities of Newtonian dynamics.

9.3.2 The Poincaré group and special relativity

Newtonian dynamics functioned well as a description of spacetime up until the late nineteenth century. At that time however, two new developments shattered its foundations. The first nail in its coffin was the famous Michelson–Morley experiment (1887), which refuted the best current explanation of the properties of light within Newtonian theory in terms of the 'theory of ether'. The simplest interpretation of their result was that the speed of light was independent of the speed of the observer, in stark contradiction with the Galilean group, which obviously cannot accommodate such behaviour. A second (closely related) fact involves Maxwell's equations of electromagnetism, which are not invariant under the Galilean group.

After an exciting decade of developments, best summarised elsewhere, Einstein's 1905 foundational paper spelled out a new theory, *special relativity*, based on a different set of principles. Four dimensional spacetime is henceforth to be modelled on $\mathbb{R}^{1,3}$, which is shorthand for a space with coordinates $\mathbf{x} = (t, x_1, x_2, x_3)$ and *Lorentz pseudometric*

$$\mathrm{d}s^2 = -c^2 \mathrm{d}t^2 + \mathrm{d}x_1^2 + \mathrm{d}x_2^2 + \mathrm{d}x_3^2;$$

or, if the infinitesimal notation is unfamiliar, you can write the *Lorentz distance* of vectors $\mathbf{x} = (t, x_i)$, $\mathbf{y} = (s, y_i) \in \mathbb{R}^{1,3}$ as

$$d(\mathbf{x}, \mathbf{y}) = -c^2(t - s)^2 + \sum_i (x_i - y_i)^2.$$

(The sign we adopt is the opposite to most physics texts.) Here the constant c, with the classical dimensions length/time, is the *speed of light*, postulated to be universal in all inertial coordinate systems. In theoretical discussions, one often sets $c = 1$ for reasons of convenience.

In special relativity, the only restriction on changes of reference frame is that the Lorentz (pseudo-)distance on $\mathbb{R}^{1,3}$ (and the 'positive light-cone') is preserved; this is *Einstein's relativity principle*. The group of such transformations is the *Poincaré group*[1] Poin(1, 3) consisting of maps

$$\mathbf{x} \mapsto A\mathbf{x} + \mathbf{b},$$

where $A \in \mathrm{O}^+(1, 3)$ is a Lorentz matrix (preserving the positive cone), and $\mathbf{b} \in \mathbb{R}^{1,3}$. This group can be studied in complete analogy with the treatment of 6.5.3: it is the semidirect product

$$\mathrm{Poin}(1, 3) \cong \mathrm{O}^+(1, 3) \rtimes \mathbb{R}^{1,3}$$

of a normal subgroup, the group $\mathbb{R}^{1,3}$ of spacetime translations, and the four dimensional Lorentz group $\mathrm{O}^+(1, 3)$. Also, for fixed values of the time variable t, the metric reduces to the Euclidean metric on a copy of $\mathbb{R}^3$. Hence Poin(1, 3) contains a subgroup Eucl(3) of Euclidean transformations. However, since the Poincaré group mixes t and x coordinates, this splitting of spacetime into 'time' and 'space' is not canonical, but depends on the choice of coordinate frame (observer).

Hyperbolic geometry is contained in the Lorentz space $\mathbb{R}^{1,n}$ of special relativity as the space-like hypersurface

$$q_L(t, x_i) = -1 \quad \text{with } t > 0.$$

---

[1] The naming of concepts during these exciting years was rather haphazard, often respecting accident and scientific standing more than historical accuracy. In particular, the so-called Lorentz metric appears to have been proposed first (albeit implicitly) by the Irish physicist George FitzGerald, followed (now explicitly) by another Irishman, Sir Joseph Larmor and only for the third time by Lorentz himself. Poincaré came very close to inventing special relativity in the years 1900–1904, showing in particular that Lorentz transformations form a group; hence in the case of the Poincaré group, the name is accurate.

The distinction of time-like and space-like vectors in the Lorentz model of hyperbolic geometry derives exactly from this physical interpretation.

9.3.3
Wigner's
classifica-
tion:
elementary
particles

As discussed above, the Poincaré group Poin(1, 3) contains the Euclidean group Eucl(3), hence also the Euclidean rotation group SO(3). As you recall from 8.5–8.6, the latter group has a double cover SU(2) → SO(3), that is, a two-to-one surjective group homomorphism with kernel ±1. It turns out that this double cover extends to a double cover

$$\widetilde{\mathrm{Poin}}(1, 3) \to \mathrm{Poin}(1, 3)$$

of the Poincaré group, which can be constructed using the group $\mathrm{SL}(2, \mathbb{C})$ of 2×2 complex matrixes of determinant 1 (which obviously contains the group SU(2) covering SO(3)).

One of the first spectacular uses of group theory in theoretical physics was Wigner's insight of the 1940s, which relates 'symmetries of spacetime' to 'things in it' (particles), and can be summarised as follows (see Sternberg [23] for the physical intuition and more details).

(1) An 'elementary particle' of nature is a (finite dimensional, irreducible, unitary) representation of the symmetry group of spacetime, satisfying certain 'physical restrictions'.

(2) The symmetry group of spacetime is the Poincaré group, or more precisely its universal cover $\widetilde{\mathrm{Poin}}(1, 3)$.

(3) The classification of the relevant representations of the Poincaré group thus leads to a classification of all elementary particles.

Recall from 8.8 that a (linear) representation of a group G is a group homomorphism from G to a group of (complex) matrixes; a unitary representation is one where the image of every element of G is a unitary matrix (the latter restriction arises from quantum mechanics, which need not unduly worry us at this point).

Wigner proved that 'physically relevant' representations of $\widetilde{\mathrm{Poin}}(1, 3)$ are classified by

- a continuous nonnegative parameter $m \geq 0$, called the *rest mass* of the particle, and
- a half-integer s, called *particle spin*, that is allowed to take nonnegative values $0, \frac{1}{2}, 1, \ldots$ for particles of mass $m > 0$, and all values $0, \pm\frac{1}{2}, \pm1, \ldots$ for those with $m = 0$.

Integral spin particles correspond to representations for which the kernel $\pm1 = \mathrm{ker}(\widetilde{\mathrm{Poin}}(1, 3) \to \mathrm{Poin}(1, 3))$ acts trivially, so really representations of Poin(1, 3); whereas for particles with half-integral spin, the double covering is necessary. Examples of the two kinds are *photons*, which are massless (that is, $m = 0$) and have integral spin $s = 1$, and electrons with $s = \frac{1}{2}$ and a certain positive value of m. (The phenomenon of spin $\frac{1}{2}$ particles was the main point of the discussion of 8.7.) The group $\widetilde{\mathrm{Poin}}(1, 3)$ has additional 'nonphysical' representations with $m^2 < 0$; these are called *tachyons* (mythical particles travelling faster than the speed of light), and are relegated to the world of science fiction in most current theories (but not all).

9.3.4
The
Standard
Model and
beyond

The importance of Wigner's insight in the development of modern physics can hardly be overstated: in a sense, it concludes another 2000 plus year old story, the search for the ultimate building blocks of the physical universe, and does so in mathematical terms. Of Wigner's program, (1) and (3) have stood as cornerstones of most theories of particle physics proposed in the last 50 years. Only (2), the specific choice of the symmetry group, has changed during the course of subsequent developments.

One thing that was clear already at the outset is that Wigner's original discussion does not incorporate the electromagnetic interactions of elementary particles. This however only requires a minor modification, taking into account an additional *internal* symmetry group U(1). This group is no longer a geometric symmetry of spacetime, but rather a symmetry of the whole theory of electromagnetism in spacetime, used to encode additional data. Representations of the combined group $\widetilde{\text{Poin}} \times \text{U}(1)$ are now parametrised by a triple of numbers (m, q, s), with the additional quantum number q, the *electric charge*, taking integer values. In fact, internal symmetry groups such as the U(1) of electromagnetism do not have to appear as a single group for the whole theory; much more powerfully, each particle can have a fibre bundle of these symmetry groups over the whole of spacetime, leading to the idea of gauge theory.

As the particle accelerators of the 1950s and 1960s grew capable of producing faster and faster particles and slamming them into one another at higher and higher energies, the zoo of known elementary particles grew accordingly. Alongside this, the internal symmetry group also changed, accommodating various features of particles to do with newly discovered forces, the *strong and weak nuclear forces* of particle physics. In Wigner style, new groups led in turn to the prediction of new particles, and their existence was in many cases confirmed in subsequent accelerator experiments. There is really no space here to elaborate on this development; I recommend Sternberg [23] as a good source. Let me only say that the most popular current theory is the *Standard Model*, based on the Poincaré group augmented by the internal symmetry group U(1) $\times$ SU(2) $\times$ SU(3); roughly, the three factors are responsible for the electromagnetic, weak and strong forces (this is of course a gross over-simplification).

Embedding the internal symmetry group U(1) $\times$ SU(2) $\times$ SU(3) into an even larger group, mixing all three forces (electromagnetic, weak and strong) completely, come under the name *Grand Unification Theory* (GUT), a sometime favourite pastime of 'armchair physics'. Popular GUT groups include the special unitary group SU(5), the group SO(10), and even more exotic constructs such as the 'exceptional' groups called E_6 and E_8. It is hard, however, for any of these exotic theories to establish a domination over their rivals; part of the problem seems to be that the Standard Model works so well, and explains to remarkable accuracy almost everything one could hope to see in experiments using accelerators of the present and near future; thus anomalous measurements against which you can check your latest GUT group are few and far between.

**9.3.5
Other
connections**

The connections between geometry and physics extend beyond the relationship between spacetime symmetries and particles. The two crowning achievements of early twentieth century physics, quantum theory and general relativity, are inextricably linked to the ideas of geometry in a number of ways. The influence of the discovery of hyperbolic geometry on relativity has already been mentioned: the fact that hyperbolic geometry has intrinsic curvature changed physical intuition, culminating in Einstein's insight that gravity, instead of acting as a classical 'force', is better described as encoded into the local curved structure of space itself (for more on this, see the next section). Quantum mechanics, invented by Schrödinger and Heisenberg in the 1920s, was axiomatised by Dirac and von Neumann, building on the Hilbert incidence axioms for projective geometry (see 5.12). Much more recently, the essential incompatibility between general relativity and quantum theory has led to the introduction and study of string theory, which builds on and generalises all of classical and modern geometry as we know it; this is however well beyond the scope of this book.

9.4   The famous trichotomy

**9.4.1
The
curvature
trichotomy
in geometry**

The metric geometries of this course come in a triad: spherical, Euclidean and hyperbolic. In terms of curvature, the three geometries correspond to the three cases of Figure 9.4a, having local curvature positive, zero or negative. You can determine which geometry you are in locally by measuring the perimeter of a circle of radius R, which, as you remember from Exercises 3.1 and 3.13, comes out to be $2\pi \sin R$, $2\pi R$ and $2\pi \sinh R$ in the three cases. The key point here is that the perimeter of a circle or the area of a disc grows exponentially with the radius in hyperbolic space, making hyperbolic space 'much bigger' than the sphere or the Euclidean plane. The curvature can also be detected by measuring the angle sum of a triangle $\triangle$ of the geometry, which is $> \pi$, equal to π and $< \pi$ in the three cases, where the excess or defect is proportional to the area of $\triangle$. Globally, as discussed at several points, the difference is visible also in the incidence properties of lines: in the sphere two lines always meet, in the Euclidean they either meet or are precisely parallel, whereas the hyperbolic plane has plenty of pairs of lines that diverge.

Topologically, the Euclidean plane $\mathbb{E}^2$, the sphere S^2 and hyperbolic space $\mathcal{H}^2$ are all simply connected (cf. 7.15; for $\mathcal{H}^2$, use the homeomorphic model $\mathcal{H}$ of Exercises 3.23–3.26 if you wish). As well as these simply connected geometries however, we can also consider compact ones; for simplicity we only discuss the oriented surfaces here. The sphere is already compact; the compact version of the plane is the one-holed torus, obtained from the plane by an equivalence relation which identifies points which are related to each other by translation by vectors in a fixed parallelogram lattice. The most exciting story is that of the hyperbolic plane, which by itself can give rise to a multitude of compact geometric spaces: it can be shown that all compact geometric surfaces with ≥ 2 holes can be derived from the hyperbolic plane (Figure 9.4b). The number of holes in a compact surface is called its *genus*; so in terms of the genus, our trichotomy becomes $g = 0$, $g = 1$ or $g > 1$. To return to the

Figure 9.4a   The cap, flat plane and Pringle's chip.

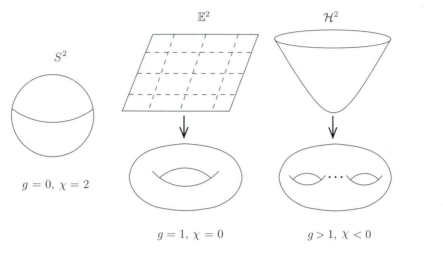

Figure 9.4b   The genus trichotomy $g = 0, g = 1, g \geq 2$ for oriented surfaces.

basic trichotomy of positive, zero or negative curvature, we can take the Euler number $\chi = 2 - 2g$ of the surface, which is simply the quantity 'faces − edges + vertexes' in Euler's formula for a triangulated surface. Then $\chi = 2$ for a sphere, as everyone knows; also $\chi = 0$ for a torus and $\chi < 0$ for the geometric surfaces with more than one hole. It is a fun exercise to triangulate a surface with two holes and check Euler's formula for it! (See Exercise 7.19 for the details.)

The classification of three dimensional geometries that extend our two dimensional curvature trichotomy rejoices in the name of *Thurston's geometrisation conjecture* (late 1970s). This includes as a humble first case the Poincaré conjecture characterising the 3-sphere; this may well turn out to be the first of the Clay Mathematical Institute's million-dollar Millennium Prize Problems to be solved. In a different direction, my own subject of classification of varieties in algebraic geometry studies geometric shapes defined in space by several polynomial equations; the curvature trichotomy reappears there in an algebraic form.

9.4.2
On the
shape and
fate of the
universe

Much was written up to the turn of the twentieth century on the subject of whether our own three dimensional universe is Euclidean, spherical or hyperbolic; Poincaré's extended essay *La science et l'hypothèse* (1902) points out that the question itself begs a number of conventions, for example on how the objects of geometry (straight lines, distance) are realised as physical objects (light rays, observations of astronomy). Maybe the answer to the question depends on our choice of conventions.

The universe has grown in size and complexity since Poincaré's day, an expansion that continues apace to this day. According to special relativity (1905), it does not make sense to consider space as a separate entity from spacetime. General relativity (1916) says that spacetime is not flat or even of constant curvature, but is curved by the presence of matter; this resolves the instantaneous action-at-a-distance that was a philosophical contradiction implicit in Newton's theory of gravitation. The existence of black holes seems to be acknowledged by the majority of astrophysicists and cosmologists, and the origin of the universe in the Big Bang some 13×10^9 years ago (give or take the odd billion years) is current orthodoxy. On a simple-minded view, these extreme events of spacetime can only be represented in geometry as singularities localised around isolated points. However, it is possible that the singularity is only in our representation, much as Mercator's projection presents a distorted view of the North pole.

A separate trichotomy concerns the long-term future of the universe – will gravity eventually slow down the expansion of the universe, causing it to collapse back on itself to a Big Crunch, so that time is also bounded in the future? will the expansion continue indefinitely, with the universe getting bigger and bigger and emptier and emptier? or are we precisely on the boundary between the two cases, so that expansion slows down to nothing? The two trichotomies are possibly logically independent, but who am I to judge?

One could believe that the general relativistic curvature effects of mass can be envisaged as merely minor localised disturbances, and that space in the large is nevertheless Euclidean; this is possibly the view held by many practising cosmologists (I have not carried out a scientific poll). However, it seems that the same population cheerfully admits that something like 80–90% of the mass of the universe is not accounted for by current theories ('black matter' and 'black energy'). Some will even admit to not having any very specially well informed view on whether spacetime is 4-dimensional or really 10- or 11-dimensional. Just a little overall curvature or cosmological constant could go a long way (compare Exercise 3.13 (c)). Given all the surprises that the study of science has brought to light in recent centuries, it might seem premature to commit oneself to an excessively firm view. There is a flourishing popular science literature on all these topics; perhaps the best informed books are those of Martin Rees, for example [20].

**9.4.3
The snack
bar at the
end of the
universe**

Even if one admits the flat and boring possibility that the universe is asymptotically Euclidean, and its expansion exactly fine tuned to slow down but never reverse, it might still happen that we get sucked into a black hole, and (who knows?) are resurrected to come out the other side as a new baby universe. At this point, you can pick and choose what you want to believe, making this a nice optimistic note on which to end my fairy story.

Appendix A Metrics

Definition   A *metric* on a set X is a specification of a *distance* $d(x, y)$ between any two points $x, y \in X$, in other words a map $d: X \times X \to \mathbb{R}$, required to satisfy the following axioms for all $x, y, z \in X$:

1. $d(x, y) \geq 0$ and $d(x, y) = 0$ if and only if $x = y$;
2. $d(x, y) = d(y, x)$;
3. the triangle inequality $d(x, y) \leq d(x, z) + d(z, y)$.

For example, the real line $\mathbb{R}$ with $d(x, y) = |x - y|$ is a metric space. The epsilon-delta definition of continuity of a function in a first calculus course uses that $\mathbb{R}$ is a metric space (compare 7.2). Theorem 1.1, Corollary 3.3 and Corollary 3.10 say that the vector space $\mathbb{R}^n$ and hence Euclidean space $\mathbb{E}^n$, the sphere S^2 and the hyperbolic plane $\mathcal{H}^2$ are all metric spaces with their respective distance functions. The set of complex numbers $\mathbb{C}$ is also a metric space under the distance function $d(z_1, z_2) = |z_2 - z_1|$. Some frivolous examples show that many distance functions in use in the real world are *not* metrics:

1. Air fares: let $d(x, y)$ be the price of an airline ticket from x to y; this is usually unsymmetric, and does not satisfy the triangle inequality.
2. The distance you travel by car to go from one point of a town to another; this is not symmetric, because of one-way traffic systems. However, it satisfies the triangle inequality, because you take the minimum over paths, at least if your taxi driver is honest.
3. For a cyclist, up a hill is of course much further than down.

I use the following simple definition to pass from a metric space to the slightly more general notion of topological space in Chapter 7 (see Section 7.2).

Definition   Let X be a metric space, $x \in X$ a point and $\varepsilon > 0$ a real number. The ball in X of radius ε centred at x is the subset

$$B(x, \varepsilon) = \{y \in X \mid d(x, y) < \varepsilon\} \subset X.$$

For example, if $X = \mathbb{R}$ is the real line, then $B(x, \varepsilon)$ is the usual open interval $(x - \varepsilon, x + \varepsilon)$. All the definitions of continuity of $f(x)$ in the first calculus course can be expressed in terms of these intervals.

Definition Let (X, d) and (Y, d_Y) be metric spaces. An *isometry* is a bijective map $f: X \to Y$ satisfying the condition

$$d_Y(f(x), f(y)) = d(x, y).$$

The meaning of this definition is that the two spaces (X, d) and (Y, d_Y) are 'the same' as far as their metric properties are concerned. An example that is used very often is the fact that the complex numbers $\mathbb{C}$ and the vector space $\mathbb{R}^2$ are isometric under the map $x + iy \mapsto (x, y)$. Note that seemingly different metric spaces can be isometric under some weird or ingenious map; see for example Exercise A.3 and, for a geometric example, Exercise 3.24.

A slightly different case of this definition that comes up all the time in geometry is when $(X, d) = (Y, d')$ and f is a bijection. Then f is viewed as a selfmap of X 'preserving all the metric geometry'. The motions of geometries studied throughout this book provide examples.

Exercises

A.1 Let X be a metric space and $t: X \to X$ a map that preserves distances $d(t(x), t(y)) = d(x, y)$. Prove that t is injective. Give an example in which t is not bijective; in other words, X can be isometric to a strict subset of itself, just as in set theory, an infinite set can be in bijection with a strict subset. [Hint: think of 'Hilbert's hotel'.]

A.2 Let $S = [1, \ldots, n]$ be a set containing n elements, and X the set of all subsets of S. For $x, y \in X$, write $d(x, y)$ for the size of the symmetric difference of x and y (the number of elements of S contained in one of x, y but not the other). Show that d is a metric on the set X. What happens to the construction if S is infinite? What happens if S is infinite but I insist that X consists only of the finite subsets of S?

A.3 Let P be the set of polynomials in one variable with coefficients in $\mathbb{Z}/2$; remember, this means that we work over the field $\{0, 1\}$ with two elements where the addition law includes $1 + 1 = 0$. If f and g are two polynomials, let $d(f, g)$ be the number of nonzero terms in the difference $f - g$. Show that d is a metric on P. Show also that P with this metric is isometric to some metric space appearing in the previous exercise.

A.4 Prove that a metric space with exactly 3 points is isometric to a subset of $\mathbb{E}^2$.

A.5 Let $X = \{A, B, C, D\}$ with $d(A, D) = 2$, but all the other distances equal to 1. Check that d is a metric. Prove that the metric space X is not isometric to any subset of $\mathbb{E}^n$ for any n. Can you realise X as a subset of a sphere S^2 of appropriate radius, with the spherical 'great circle' metric? [Hint: I am sure you know the riddle: an explorer starts out from base camp, walks 10 miles due South, meets a bear, runs 10 miles due West, then 10 miles due North and finds himself back at base camp. *What colour was the bear?* If in doubt, turn to Figure A.1.]

Figure A.1   The bear.

Appendix B Linear algebra

The distance function in $\mathbb{R}^n$ is given by the norm $|\mathbf{x}|^2 = \sum x_i^2$, which comes from the standard inner product $\mathbf{x} \cdot \mathbf{y} = \sum x_i y_i$. The ideas here are familiar from Pythagoras' theorem and the equations of conics in plane geometry, and from the vector manipulations in $\mathbb{R}^3$ used in applied math courses. A quadratic form in variables $x_1, \ldots, x_n$ is simply a *homogeneous* quadratic function in the obvious sense. For clarity I recall the formal definitions and results from linear algebra.

B.1 Bilinear form and quadratic form

Definition Let V be a finite dimensional vector space over $\mathbb{R}$. A *symmetric bilinear form* φ on V is a map $\varphi \colon V \times V \to \mathbb{R}$ such that

(i) φ is linear in each of the two arguments, that is

$$\varphi(\lambda \mathbf{u} + \mu \mathbf{v}, \mathbf{w}) = \lambda \varphi(\mathbf{u}, \mathbf{w}) + \mu \varphi(\mathbf{v}, \mathbf{w})$$

for all $\mathbf{u}, \mathbf{v}, \mathbf{w} \in V$, $\lambda, \mu \in \mathbb{R}$, and similarly for the second argument,

(ii) $\varphi(\mathbf{u}, \mathbf{v}) = \varphi(\mathbf{v}, \mathbf{u})$ for all $\mathbf{u}, \mathbf{v} \in V$.

A *quadratic form* q on V is a map $q \colon V \to \mathbb{R}$ such that

$$q(\lambda \mathbf{u} + \mu \mathbf{v}) = \lambda^2 q(\mathbf{u}) + 2\lambda\mu\varphi(\mathbf{u}, \mathbf{v}) + \mu^2 q(\mathbf{v})$$

for all $\mathbf{u}, \mathbf{v} \in V$, $\lambda, \mu \in \mathbb{R}$, where $\varphi(\mathbf{u}, \mathbf{v})$ is a symmetric bilinear form.

Proposition *A quadratic form is determined by a symmetric bilinear form and vice versa by the rules*

$$q(x) = \varphi(\mathbf{x}, \mathbf{x}) \quad \text{and} \quad \varphi(\mathbf{x}, \mathbf{y}) = \frac{1}{2}\big(q(\mathbf{x} + \mathbf{y}) - q(\mathbf{x}) - q(\mathbf{y})\big).$$

Choosing a basis $\mathbf{e}_1, \ldots, \mathbf{e}_n$ of V, a quadratic form q or its associated symmetric bilinear form φ are given by

$$q(\mathbf{x}) = \sum_{i,j} a_{ij} x_i x_j = {}^t\mathbf{x} K \mathbf{x}, \quad \varphi(\mathbf{x}, \mathbf{y}) = \sum_{i,j} a_{ij} x_i y_j = {}^t\mathbf{x} K \mathbf{y}.$$

Here $\mathbf{x} = {}^t(x_1, \ldots, x_n) = \sum x_i \mathbf{e}_i$, $\mathbf{y} = {}^t(y_1, \ldots, y_n) = \sum y_i \mathbf{e}_i$ and $K = (k_{ij})$ is a symmetric matrix whose entries are given by $k_{ij} = \varphi(\mathbf{e}_i, \mathbf{e}_j)$.

B.2   Euclid and Lorentz

There are two special bilinear forms that are useful in geometry. To see the first, let $V = \mathbb{R}^n$ be the vector space with the standard basis

$$\mathbf{e}_1 = {}^t(1, 0, \ldots, 0), \ldots, \mathbf{e}_n = {}^t(0, \ldots, 0, 1).$$

The *Euclidean inner product* corresponds to the matrix

$$I = \operatorname{diag}(1, 1, \ldots, 1).$$

It is the familiar

$$\varphi_E(\mathbf{x}, \mathbf{y}) = \mathbf{x} \cdot \mathbf{y} = {}^t\mathbf{x} I \mathbf{y} = \sum_i x_i y_i,$$

with corresponding quadratic form

$$q_E(\mathbf{x}) = |\mathbf{x}|^2 = \sum_i x_i^2.$$

As you know, an *orthonormal basis* of $\mathbb{R}^n$ is a set of n vectors $\mathbf{f}_1, \ldots, \mathbf{f}_n \in \mathbb{R}^n$ such that

$$\mathbf{f}_i \cdot \mathbf{f}_j = \delta_{ij} = \begin{cases} 0 & \text{for } i \neq j \\ 1 & \text{for } i = j. \end{cases}$$

The model for this definition is the usual basis $\mathbf{e}_i = (0, \ldots, 1, 0, \ldots)$ of $\mathbb{R}^n$ (with 1 in the ith place). The inner product φ_E expressed in terms of an orthonormal basis $\mathbf{f}_1, \ldots, \mathbf{f}_n$ of V still has matrix I.

For the indefinite case, it is convenient to change notation slightly, so let $V = \mathbb{R}^{n+1}$ be the vector space with the standard basis $\mathbf{e}_0, \ldots, \mathbf{e}_n$. The *Lorentz dot product* is the symmetric bilinear form given by the matrix

$$J = \operatorname{diag}(-1, 1, \ldots, 1).$$

If $\mathbf{x} = (t, x_1, \ldots, x_n)$ and $\mathbf{y} = (s, y_1, \ldots, y_n)$ then

$$\varphi_L(\mathbf{x}, \mathbf{y}) = (t, x_1, \ldots, x_n) \cdot_L (s, y_1, \ldots, y_n) = -ts + \sum x_i y_i.$$

The *Lorentz norm* is the associated quadratic form $q_L : V \to \mathbb{R}$, defined by

$$q_L(t, x_1, \ldots, x_n) = -t^2 + \sum x_i^2.$$

A *Lorentz basis* $\mathbf{f}_0, \mathbf{f}_1, \ldots, \mathbf{f}_n$ is a basis of V as a vector space, with respect to which q_L has the standard diagonal matrix J; that is,

$$q_L(\mathbf{f}_0) = -1, \quad q_L(\mathbf{f}_i) = 1 \quad \text{for } i \geq 1 \quad \text{and} \quad \mathbf{f}_i \cdot_L \mathbf{f}_j = 0 \quad \text{for } i \neq j.$$

B.3 Complements and bases

Let (V, φ) be a vector space with bilinear form.

Definition For a vector subspace $W \subset V$, define the *complement* of W with respect to φ to be

$$W^\perp = \left\{ \mathbf{x} \in V \mid \varphi(\mathbf{x}, \mathbf{w}) = 0 \quad \text{for all } \mathbf{w} \in W \right\}.$$

In general, complements need not have any particularly nice properties; notice for example that the zero inner product (with matrix $K = 0$) gives $W^\perp = V$ for all subspaces W. However, for 'nice' inner products the situation is completely different. I write this section explicitly with the minimal generality needed for the geometric applications; all this can be souped up to obtain the general Gram–Schmidt process, Sylvester's law of inertia, etc.

Theorem *Let φ be the Euclidean inner product on $V = \mathbb{R}^n$. Let W be a subspace of $\mathbb{R}^n$. Then*

(1) W *has an orthonormal basis* $\mathbf{f}_1, \ldots, \mathbf{f}_k$,
(2) *any vector* $\mathbf{v} \in \mathbb{R}^n$ *has a unique expression* $\mathbf{v} = \mathbf{w} + \mathbf{u}$ *with* $\mathbf{w} \in W$ *and* $\mathbf{u} \in W^\perp$; *in other words,* $\mathbb{R}^n$ *is the direct sum* $W \oplus W^\perp$.

Proof Suppose that W is not the zero vector space, take a nonzero $\mathbf{v}_1 \in V$ and let $\mathbf{f}_1 = \mathbf{v}_1/|\mathbf{v}_1|$ be a vector with unit length in the direction of $\mathbf{v}_1$. If $\mathbf{f}_1$ spans W then I am home. If not, take $\mathbf{v}_2$ outside the span of $\mathbf{f}_1$ and let $\mathbf{f}_2$ be a unit vector in the direction of $\mathbf{v}_2 - (\mathbf{v}_2 \cdot \mathbf{f}_1)\mathbf{f}_2$. Then, as you can check, the cunning choice of the direction of $\mathbf{f}_2$ ensures that it is orthogonal to $\mathbf{f}_1$, and it lies in W. Now continue this way by induction. Either the constructed $\mathbf{f}_1, \ldots, \mathbf{f}_k$ generate W, or you can find $\mathbf{v}_{k+1} \in W$ outside their span, and then a unit vector in the direction of $\mathbf{v}_{k+1} - \sum(\mathbf{v}_{k+1} \cdot \mathbf{f}_i)\mathbf{f}_i$ can be added to the collection.

For the second statement, find an orthonormal basis $\mathbf{f}_1, \ldots, \mathbf{f}_k$ of W, and extend it using the same method to an orthonormal basis $\mathbf{f}_1, \ldots, \mathbf{f}_n$ of $\mathbb{R}^n$. Then every vector $v \in \mathbb{R}^n$ has a unique expression

$$\mathbf{v} = \sum_{i=1}^n \lambda_i \mathbf{f}_i$$

and then

$$\mathbf{w} = \sum_{i=1}^k \lambda_i \mathbf{f}_i, \quad \mathbf{u} = \sum_{i=k+1}^n \lambda_i \mathbf{f}_i$$

is the only possible choice. QED

The procedure of the proof is algorithmic, so lends itself easily to calculations; to make sure that you understand it, do Exercise B.1.

Theorem   Let $V = \mathbb{R}^{n+1}$ with the Lorentz dot product and form.

(3)   Let $\mathbf{v} \in \mathbb{R}^{n+1}$ be any vector with $q_L(\mathbf{v}) < 0$. Then $q_L(\mathbf{w}, \mathbf{w}) > 0$ for $\mathbf{w}$ a nonzero vector in the Lorentz complement $\mathbf{v}^{\perp}$.

(4)   Let $\mathbf{f}_0 \in \mathbb{R}^{n+1}$ be a vector with $q_L(\mathbf{f}_0) = -1$. Then $\mathbf{f}_0$ is part of a Lorentz basis $\mathbf{f}_0, \dots, \mathbf{f}_n$ of $\mathbb{R}^{n+1}$.

Proof   For (3), suppose that $\mathbf{v} = (t, x_1, \dots, x_n)$ and $\mathbf{w} = (s, y_1, \dots, y_n)$ satisfy $q_L(\mathbf{v}) < 0$ and $\mathbf{v} \cdot_L \mathbf{w} = 0$, that is

$$-t^2 + \sum_{i=1}^{n} x_i^2 < 0 \tag{1}$$

and

$$-st + \sum_{i=1}^{n} x_i y_i = 0. \tag{2}$$

Then (1) and (2) give that

$$\left(-s^2 + \sum_{i=1}^{n} y_i^2\right)t^2 = -s^2 t^2 + t^2\left(\sum_{i=1}^{n} y_i^2\right)$$

$$> -\left(\sum_{i=1}^{n} x_i y_i\right)^2 + \left(\sum_{i=1}^{n} x_i^2\right)\left(\sum_{i=1}^{n} y_i^2\right),$$

provided that the y_i are not all 0. But we know that the last line is ≥ 0 (in fact it is equal to $\sum(x_i y_j - x_j y_i)^2$, compare 1.1), so

$$-s^2 + \sum_{i=1}^{n} y_i^2 > 0$$

which is the statement.

For (4), pick $\mathbf{v}_1 \in \mathbb{R}^{n+1}$ linearly independent of $\mathbf{f}_0$ and set

$$\mathbf{w}_1 = \mathbf{v}_1 + (\mathbf{f}_0 \cdot_L \mathbf{v}_1)\mathbf{f}_0.$$

Then $\mathbf{w}_1$ is a nonzero element of $\mathbf{f}_0^{\perp}$, so by (3) it has positive Lorentz norm. Hence I can set $\mathbf{f}_1 = \mathbf{v}_1/\sqrt{q_L(\mathbf{v}_1)}$. Then by construction $\mathbf{f}_0, \mathbf{f}_1$ are part of a Lorentz basis. Now continue with the inductive method used in the proof of the previous theorem.   QED

B.4   Symmetries

Return to the case of a general symmetric bilinear form φ on the vector space V, and its associated quadratic form q.

Proposition   *Let $\alpha : V \to V$ be a linear map. Then equivalent conditions:*

1. *α preserves q, that is, $q(\alpha(\mathbf{x})) = q(\mathbf{x})$ for all $\mathbf{x} \in V$,*
2. *α preserves φ, that is, $\varphi(\alpha(\mathbf{x}), \alpha(\mathbf{y})) = \varphi(\mathbf{x}, \mathbf{y})$ for all $\mathbf{x}, \mathbf{y} \in V$.*

Proof   The equivalence simply follows from the fact that q is determined by φ and conversely, φ is determined by q from Proposition B.1.   QED

Now identify V with $\mathbb{R}^n$ using the standard basis $\mathbf{e}_1, \ldots, \mathbf{e}_n$. Let $K = \{\varphi(e_i, e_j)\}$ be the matrix of φ.

Proposition (continued)   *Let A be the $n \times n$ matrix representing α in the given basis. Then the previous two conditions are also equivalent to*

3. *A satisfies the matrix equality ${}^t\!AKA = K$.*

Proof   Recall $\varphi(\mathbf{x}, \mathbf{y}) = {}^t\!\mathbf{x}A\mathbf{y}$. Hence

$$\varphi(\alpha(\mathbf{x}), \alpha(\mathbf{y})) = \varphi(\mathbf{x}, \mathbf{y}) \iff {}^t(A\mathbf{x})K(A\mathbf{y}) = {}^t\!\mathbf{x}\,{}^t\!AK\,A\mathbf{y} = {}^t\!\mathbf{x}K\mathbf{y}$$

and the latter holds for all $\mathbf{x}$ and $\mathbf{y}$ if and only if ${}^t\!AKA = K$.   QED

A useful observation is the following.

Lemma   *If $\det K \neq 0$ (we say that the form φ is nondegenerate) then the equivalent conditions above imply $\det A = \pm 1$.*

Proof   From (3) and properties of the determinant it follows that

$$(\det A)^2 \det K = \det K.$$

If $\det K \neq 0$ then I can divide by it.   QED

B.5 Orthogonal and Lorentz matrixes

Consider $\mathbb{R}^n$ with the Euclidean inner product, and let $\mathbf{e}_1, \ldots, \mathbf{e}_n$ with $\mathbf{e}_i = (0, \ldots, 1, 0, \ldots)$ be the usual basis. If $\mathbf{f}_1, \ldots, \mathbf{f}_n \in \mathbb{R}^n$ are any n vectors, there is a unique linear map $\alpha : \mathbb{R}^n \to \mathbb{R}^n$ such that $\alpha(\mathbf{e}_i) = \mathbf{f}_i$ for $i = 1, \ldots, n$. Namely write $\mathbf{f}_j$ as the column vector $\mathbf{f}_j = (a_{ij})$; then α is given by the matrix $A = (a_{ij})$ with columns the vectors $\mathbf{f}_j$. Now, by Proposition B.4 and by direct inspection, the following conditions are equivalent:

1. $\mathbf{f}_1, \ldots, \mathbf{f}_n$ is an orthonormal basis;
2. the columns of A form an orthonormal basis;
3. ${}^t\!AA = I$;
4. α preserves the Euclidean inner product.

We say that α is an *orthogonal transformation* and A an *orthogonal matrix* if these conditions hold. We get the following result.

Proposition $\alpha \mapsto (\alpha(\mathbf{e}_1), \ldots, \alpha(\mathbf{e}_n))$ *establishes a one-to-one correspondence*

$$\left\{ \begin{array}{c} orthogonal\ transformations \\ \alpha\ of\ \mathbb{R}^n \end{array} \right\} \leftrightarrow \left\{ \begin{array}{c} orthonormal\ bases \\ \mathbf{f}_1, \ldots, \mathbf{f}_n \in \mathbb{R}^n \end{array} \right\}.$$

If (V, φ) is Lorentz, a matrix A satisfying the condition ${}^t\!AJA = J$ of Proposition B.4 (3) is called a *Lorentz matrix*. I leave you to formulate the analogous correspondence between Lorentz bases and Lorentz matrixes.

B.6   Hermitian forms and unitary matrixes

This section discusses a slight variant of the above material, for vector spaces over the field $\mathbb{C}$ of complex numbers. Let V be a finite dimensional vector space over $\mathbb{C}$. A *Hermitian form* $\varphi \colon V \times V \to \mathbb{C}$ is a map satisfying the conditions

$$\varphi(\lambda\mathbf{u} + \mu\mathbf{v}, \mathbf{w}) = \overline{\lambda}\varphi(\mathbf{u}, \mathbf{w}) + \overline{\mu}\varphi(\mathbf{v}, \mathbf{w})$$

and

$$\varphi(\mathbf{u}, \lambda\mathbf{v} + \mu\mathbf{w}) = \lambda\varphi(\mathbf{u}, \mathbf{v}) + \mu\varphi(\mathbf{u}, \mathbf{w}),$$

where $\lambda, \mu \in \mathbb{C}$; note the appearance of the complex conjugate in the first row. The corresponding *Hermitian norm* q on V is

$$q(\mathbf{v}) = \varphi(\mathbf{v}, \mathbf{v}).$$

The relation between φ and q is slightly more complicated than in the real case; I leave you to check the rather daunting looking identity

$$\varphi(\mathbf{u}, \mathbf{v}) = \frac{1}{4}\Big(q(\mathbf{u} + \mathbf{v}) - q(\mathbf{u} - \mathbf{v}) + iq(\mathbf{u} + i\mathbf{v}) - iq(\mathbf{u} - i\mathbf{v})\Big).$$

The terms in the identity are not so important; what is important is the fact that q gives back φ.

Since I am only interested in a special case, I choose a basis $\{\mathbf{e}_1, \ldots, \mathbf{e}_n\}$ of V straight away and assume that

$$\varphi(\lambda_1\mathbf{e}_1 + \cdots + \lambda_n\mathbf{e}_n, \mu_1\mathbf{e}_1 + \cdots + \mu_n\mathbf{e}_n) = \overline{\lambda}_1\mu_1 + \cdots + \overline{\lambda}_n\mu_n.$$

Such a form is called a *definite* Hermitian form. Under φ, $\mathbf{e}_1, \ldots, \mathbf{e}_n$ form a Hermitian or orthonormal basis: $\varphi(\mathbf{e}_i, \mathbf{e}_j) = \delta_{ij}$.

The following is completely analogous to Proposition B.4.

Proposition   *Let $\alpha \colon V \to V$ be a linear map represented by the $n \times n$ matrix A in the given basis. Then the following are equivalent:*

1. α *preserves the norm q;*
2. α *preserves the Hermitian form φ;*

3. A satisfies $^{\mathrm{h}}A\,A = I_n$, where $^{\mathrm{h}}A$ is the Hermitian conjugate *defined by* $^{\mathrm{h}}A = \overline{{}^t A}$; *that is*, $(^{\mathrm{h}}A)_{ij} = \overline{A}_{ji}$.

The transformation α or the matrix A representing it is *unitary* if it satisfies these conditions; the set of $n \times n$ unitary matrixes is denoted $\mathrm{U}(n)$. Unitary transformations (possibly on infinite dimensional spaces) have many pleasant properties which makes them ubiquitous in mathematics. They are also the basic building blocks of quantum mechanics and hence presumably nature; in this book I discuss one tiny example of this in 8.7.

Exercises

B.1   Let $\mathbf{f}_1 = (2/3, 1/3, 2/3)$ and $\mathbf{f}_2 = (1/3, 2/3, -2/3) \in \mathbb{R}^3$; find all vectors $\mathbf{f}_3 \in \mathbb{R}^3$ for which $\mathbf{f}_1, \mathbf{f}_2, \mathbf{f}_3$ is an orthonormal basis.

B.2   By writing down explicitly the conditions for a 2×2 matrix to be Lorentz, show that any such matrix has the form

$$\begin{pmatrix} \cosh s & \sinh s \\ \sinh s & \cosh s \end{pmatrix} \quad \text{or} \quad \begin{pmatrix} \cosh s & -\sinh s \\ \sinh s & -\cosh s \end{pmatrix}.$$

B.3   This exercise is a generalisation of the previous one; it shows that any Lorentz matrix can be put in a simple normal form in a suitable Lorentz basis; the Euclidean case is included in the main text in 1.11. Let $\alpha\colon \mathbb{R}^{n+1} \to \mathbb{R}^{n+1}$ be a linear map given by a Lorentz matrix A. Prove that there exists a Lorentz basis of $\mathbb{R}^{n+1}$ in which the matrix of α is

$$B = \begin{pmatrix} \pm 1 & & & & & \\ & I_{k^+} & & & & \\ & & -I_{k^-} & & & \\ & & & B_1 & & \\ & & & & \ddots & \\ & & & & & B_l \end{pmatrix} \quad \text{or} \quad B = \begin{pmatrix} B_0 & & & & & \\ & I_{k^+} & & & & \\ & & -I_{k^-} & & & \\ & & & B_1 & & \\ & & & & \ddots & \\ & & & & & B_{l-1} \end{pmatrix}$$

where $B_0 = \pm \left(\begin{smallmatrix} \cosh\theta_0 & \sinh\theta_0 \\ \sinh\theta_0 & \cosh\theta_0 \end{smallmatrix} \right)$, $B_i = \left(\begin{smallmatrix} \cos\theta_i & -\sin\theta_i \\ \sin\theta_i & \cos\theta_i \end{smallmatrix} \right)$ for $i > 0$, and $I_{k^\pm}$ are identity matrixes. [Hint: argue as in the Euclidean case in 1.11.2; the only extra complication is that you have to take into account the sign of the Lorentz form on the eigenvectors. The statement follows by sorting out the cases that can arise.]

B.4   Prove that a unitary matrix has determinant $\det A \in \mathbb{C}$ of absolute value 1.

References

[1]   Michael Artin, *Algebra*, Englewood Cliffs, NJ: Prentice Hall, 1991.

[2]   Alan F. Beardon, *The Geometry of Discrete Groups*, New York: Springer, 1983.

[3]   Roberto Bonola, *Non-Euclidean Geometry: a Critical and Historical Study of its Developments*, New York: Dover, 1955.

[4]   J. H. Conway and D. A. Smith, *On Quaternions and Octonions*, Natick, MA: A. K. Peters, 2002.

[5]   H. S. M. Coxeter, *Introduction to Geometry*, 2nd edn, New York: Wiley, 1969.

[6]   Peter H. Dana, The Geographer's Craft Project 1999, http://www.colorado.edu/geography/ gcraft/notes/mapproj/mapproj.html.

[7]   Richard P. Feynman, *The Feynman Lectures on Physics, Vol. 3: Quantum Mechanics*, Reading, MA: Addison-Wesley, 1965.

[8]   C. M. R. Fowler, *The Solid Earth*, Cambridge: Cambridge University Press, 1990.

[9]   William Fulton and Joseph Harris, *Representation Theory, a First Course*, Readings in Mathematics, New York: Springer, 1991.

[10]   James A. Green, *Sets and Groups, a First Course in Algebra*, London: Chapman and Hall, 1995.

[11]   Marvin J. Greenberg, *Euclidean and non-Euclidean Geometries: Development and History*, 3rd edn, New York: W. H. Freeman, 1993.

[12]   Robin Hartshorne, *Geometry: Euclid and Beyond*, Undergraduate Texts in Mathematics, New York: Springer, 2000.

[13]   David Hilbert, *Foundations of Geometry*, 2nd edn, LaSalle: Open Court, 1971.

[14]   Walter Ledermann, *Introduction to the Theory of Finite Groups*, Edinburgh: Oliver and Boyd, 1964.

[15]   Pertti Lounesto, *Clifford Algebras and Spinors*, Cambridge: Cambridge University Press, 1997.

[16]   Dana Mackenzie, A sine on the road to Mecca, *American Scientist*, **89** (3) (May–June 2001).

[17]   P. M. Neumann, G. A. Story and E. C. Thompson, *Groups and Geometry*, Oxford: Oxford University Press, 1994.

[18]   V. V. Nikulin and I. R. Shafarevich, *Geometries and Groups*, Berlin: Springer Universitext, 1987.

[19]   Elmer Rees, *Notes on Geometry*, Berlin: Springer, 1983.

[20]   Martin Rees, *Before the Beginning*, Simon and Schuster, 1997.

[21]   Walter Rudin, *Principles of Mathematical Analysis*, 3rd edn, New York: McGraw-Hill, 1976.

[22]   Graeme Segal, *Lie groups*, in R. Carter, G. Segal and I. G. Macdonald, *Lectures on Lie groups and Lie algebras*, CUP/LMS student texts, Cambridge: Cambridge University Press, 1995.

[23]   Shlomo Sternberg, *Group Theory and Physics*, Cambridge: Cambridge University Press, 1994.

[24]   W. A. Sutherland, *Introduction to Metric and Topological Spaces*, Oxford: Clarendon Press, 1975.

Index

abstract group, 169
affine
  frame, 69, 71
  geometry, 62–72, 95
  group Aff(n), 102, 161, 170
  linear
    dependence, 68, 71
    map, 8–9, 27, 68–69
    subspace, 29–30, 62–68, 70–72, 91
  space $\mathbb{A}^n$, 62, 63, 68, 95, 170
    in projective space, 82
  span, 62, 66–67
  transformation, xvi, 8–9, 68–70, 91
algebraic topology, xv, 113, 130
algebraically closed field, 136–137
angle, 1, 5–6, 27, 62, 69, 95
  bisector, 23, 25
  of rotation, 15–18
  signed, 6
  sum, 19–20, 34, 40, 51–56
angular
  defect, excess, see *angle sum*
  momentum, 93, 154
area, 40–41, 51–56
associative law, 28, 32, 94, 169
axiomatic projective geometry, 86–88, 164, 168,
    177

ball, 58, 109, 138, 146
based loop, 131–133, 136–137
basis for a topology, 124–126
bilinear form, see *Euclidean inner product,*
    Lorentz dot product, 162, 183–185
Bolyai's letter, 166

centre of rotation, 15
centroid, 21, 69–71
circumcentre, 21, 22
closed, see *compact versus closed*, 58, 75, 108,
    111, 113, 138, 148

and bounded, 115–129, 146
  diagonal, 127–128
  map, 129–130
cofinite topology, 108, 111, 127
commutative law, 15, 17, 28, 32
compact, see *maximal –, sequentially –*, xv, 75,
    115–117, 121, 133–138, 143, 146,
    152
  Lie group, 146, 147, 160
  surface, 119, 177–178
  versus closed, 128–129
compactification, 75
complex number, 12, 27, 136, 188
composite
  of maps, 26–33
  of reflections, 16, 29–31, 33, 58
  of rotation and glide, 33
  of rotation and reflection, 31
  of rotations, 27, 33
  of translations, 27
congruent triangles, 19, 25, 55
connected, see *path –, simply –*, 113–115, 117,
    138, 148, 149, 152, 153
  component, 114–115, 122, 144, 148, 149, 153,
    160, 161
  Lie group, 160
continuous, xv, 5, 68, 91, 100, 142–144, 148,
    149
  family of paths, 131–132
contractible loop, 130–133, 136,
    141
coordinate
  changes, xiv
  frame, xiv, 1
  geometry, xiii, xvi, 168
  system, xiv, 4
Coventry market, 92–93
cross-ratio, 79–81, 90, 106
curvature, 34, 40, 49, 93, 167, 177, 178,
    182

Desargues' theorem, 82–84, 88, 90
dimension, 66, 67, 70, 76, 144, 145, 160
　of a Lie group, 144–146, 148, 161
　of intersection, 67, 69, 72–73, 77, 81, 83, 88
direct motion, 10, 15, 17, 148, 151–152
disc, 111, 122, 130, 133, 139
discrete topology, 108, 110, 127, 143
distance, see *Euclidean –, hyperbolic –, metric,*
　shortest –, spherical –
　function, 1, 2, 4, 6, 7, 35, 62, 95, 180, 181, 183
duality, 85–86, 90

Einstein's
　field equations, see *general relativity*, 93
　relativity principle, see *special relativity*, 174
electron, xvi, 143, 154–159, 175
empty set, 68, 70, 72, 73, 76, 108, 124
Erlangen program, xiv–xv, 95–96, 112, 170–171
Euclid's postulates, see *parallel postulate*,
　165–167
Euclidean
　angle, 45
　distance, 1, 2, 4, 116, 151
　frame, 1, 14, 25, 40, 145
　geometry, 4, 19, 25, 34, 45, 47, 69, 95, 166
　group Eucl(n), xvi, 159, 161
　inner product, 2, 5, 9, 24, 43, 58, 184, 185, 187
　line, 4
　motion, see *motion*, 9, 10, 14, 24, 25, 47, 92,
　　144
　plane $\mathbb{E}^2$, 6, 33
　space $\mathbb{E}^n$, 1, 4–10, 29, 35, 180
　translation, 19
Euler number, 140, 177

family of paths, 131
Feuerbach circle, 23
frame, see *affine –, coordinate –, Euclidean –,*
　orthogonal –, projective –, spherical –
frame of reference, see *projective frame*
fundamental
　group, xv, 113, 130, 159
　theorem of algebra, 136

Galilean group, 93, 172–173
general
　linear group GL(n), xv, 95, 99, 101, 105, 124,
　　143, 145, 147, 148, 160, 161, 171
　relativity, 93, 167, 176, 178
generators, 29, 100–101, 103, 106
genus, 120, 139, 177
geodesic, see *shortest distance*
glide, 15–17, 24, 31–33, 40, 47, 98
　reflection, see *glide*
glueing, see *quotient topology*
great circle, see *spherical line*

group, see *abstract –, fundamental –, Galilean –,*
　general linear –, Lie –, Lorentz –,
　Poincaré –, projective linear –,
　reflection –, rotation –, spinor –,
　topological –, transformation –, unitary –

half-turn, 12, 32
Hausdorff, 109, 110, 127–130, 139, 152
Heine–Borel theorem, 116
Hermitian form, 153, 156, 160, 163, 188
homeomorphism, 107, 111, 113, 117, 119–121,
　130, 132, 134–136, 138, 139, 147, 149,
　152, 153, 160, 177
　criterion, 111, 130, 142, 152
　problem, xv, 113
homogeneous space, 169–170
hyperbolic
　distance, 43, 46, 58
　geometry, 4, 20, 34, 36, 41–167
　line, 43, 46–50, 60
　motion, 46, 144
　plane $\mathcal{H}^2$, 39, 47–49, 58–61, 180
　sine rule, 59
　space, 35, 42, 51, 104
　translation, 47, 58, 61
　triangle, 44, 51, 58, 59
　trig, 35, 44–45
hyperplane, 29, 30, 66, 67, 76, 78, 81, 82, 89, 96
　at infinity, 88

ideal point, see *infinity, point at*
ideal triangle, 51, 53–56
incentre, 23, 25
incidence of lines, 34, 40, 47, 69, 84
indiscrete topology, 108, 111, 139
infinity
　hyperplane at, 72, 73, 76, 82, 90
　point at, 48–49, 51, 53, 55, 59, 73, 75, 76,
　　79
intersection, see *dimension of –*, 108
intrinsic
　curvature, 34, 40, 177
　distance, 40
　unit, 34, 49
isometry, see *motion, preserves distances*, 4, 6,
　112, 181

Klein bottle, xiv, 139

length of path, 5
Lie group, see *compact –*, 142–164, 169
line, 4, 65
　hyperbolic, 44
　segment, 3, 65
　spherical, 35
loop, 107–137, 140, 159

Lorentz
  basis, 44, 55, 185, 186, 188, 189
  complement, 48, 186
  dot product $\cdot_L$, 43, 184, 186
  form q_L, 42, 47, 184, 186, 189
  group, 93, 159, 161
  matrix, 42, 46–47, 161, 188, 189
  norm, 44, 184, 186
  orthogonal, 44
    matrix, 187
  pseudometric, 42, 58, 174
  reflection, 47
  space, 42, 46, 188
  transformation, 47, 54, 92, 144
  translation, see *hyperbolic –*

maximal compact subgroup, 160
Mercator's projection, 139, 164, 179
metric, 180–182
  geometry, 64, 177
  space, 1, 4, 38, 180–182
  topology, 109, 125, 143, 152
minimum over paths, 5, 180
Möbius strip, xiv, 107, 118–119, 122, 139
motion, xiv, 1, 6, 7, 9–11, 14–19, 24–26, 28–34,
    38–40, 46, 47, 58, 61, 93, 95, 97, 98, 100,
    103, 105, 106, 144, 149, 151, 152, 154,
    158, 161
mousetrap topology, 122–123
Musée Grévin, 103, 105

Newtonian dynamics, 93, 161, 172–173
non-Euclidean geometry, 34–61, 167
normal form of a matrix, 10–13, 18, 29, 98–99,
    148, 189

open set, 108–111, 113–115, 117, 118, 121, 125,
    143, 148
opposite motion, 10, 15, 17, 148
orthocentre, 22–23
orthogonal, see *Lorentz –*
  axes, 1
  complement $V^\perp$, 13, 47, 145, 171, 185
  direct sum, 151
  frame, 39
  group O(n), 144–152
  line, 158
  magnetic field, 154, 158
  matrix, 7, 9–13, 24, 29, 39, 99, 144, 146–149,
    159, 187
  plane, 29
  transformation, 9, 92, 99, 187
  vector, 5, 29, 37, 151, 162, 185

Pappus' theorem, 84–85, 88, 90
parallel

axes, 31
hyperplanes, 17, 64, 66, 67
lines, 15–17, 20–23, 27, 34, 40, 49, 62, 68, 70,
    73, 82, 166
mirrors, 103
postulate, 20, 49, 60, 166
sides, 31
vector, 16, 96
path, see *length of path, minimum over paths*, 114,
    131, 159
  connected, 114, 120, 132, 141, 149
perpendicular bisector, 16, 21, 22, 24, 29, 30,
    57
perspective, 73, 74, 81–83, 88, 90
physics, xv, xvi, 93, 160, 172–179
Poincaré group, 173–176
point at infinity, see *infinity, point at*
preserves distances, 6–7, 24, 39, 181
principal homogeneous space, see *torsor*
Pringle's potato chip, 58, 178
product topology, 126–127, 139, 143
profinite topology, 125, 126
projective
  frame, 78, 79, 90, 106, 146
  geometry, 72–91
  linear group PGL(n), 77, 95, 105, 106, 144,
    146, 171
  linear subspace, 73–77
punctured disc D^*, 120, 130, 133, 136

quadratic form, 5, 9, 42, 123, 150, 151, 183
quaternions, 149–152
quotient topology, 110, 117–119, 121–125,
    139–140, 144, 152

reflection, 1, 11, 15–17, 24, 27–30, 33, 34, 40, 58,
    103, 105
  group, 103–105
  matrix, 7, 10, 24, 42, 144
relativity, see *special –, general –*, 161
rigid body motion, see *motion*
rotary reflection, 33, 40
rotation, 1, 11, 15–18, 24, 25, 27, 29, 31–34, 39,
    40, 47, 97, 100, 103, 142, 143, 149–152,
    154, 158, 161
  group, 152
  matrix, 7, 10, 42, 144
rubber-sheet geometry, xiv, 107

sequentially compact, 115–116, 138
shortest distance, see *minimum over paths*, 4, 5,
    40, 46, 58
similar triangles, 21–23
simplex of reference, see *projective frame*
simply connected, 130, 132, 146, 160
spacetime, 93, 172–176, 178, 179

special
  linear group SL(n), 159, 175
  orthogonal group SO(n), 149, 152
  relativity, xv, 93, 144, 173–174, 178
  unitary group SU(n), 153, 176
sphere S^2, 35, 36, 39, 40, 43, 56, 58, 113, 180, 181
sphere S^n, 57, 58, 116, 121, 122, 145, 151
spherical
  disc, 56
  distance, 36–38, 40, 56, 116
  frame, 34, 40
  geometry, 4, 20, 34–41, 45, 56, 57, 164, 167,
    182
  line, 39, 40
  motion, 38, 39
  triangle, 37–38, 40, 41, 57, 182
  trig, 37, 167
spin, 143, 154, 155
spinor group Spin(n), 153, 159
Standard Model, 176
subspace topology, 117, 121, 128, 144, 147, 152
symmetry, 92–95, 160, 164, 169, 173–176

topological
  group, 143–144, 159
  property, xv, 113, 127, 131, 136, 167

topology, 94, 107–141, 143
  of $\mathbb{P}^n$, 90, 121, 139
  of SO(3), 142, 143, 149
  of S^3, 152
torsor, 169–170
torus, 119, 120, 139, 177, 178
transformation group, 26–33, 92, 94–96, 101, 104,
    112, 142–163
translation, 1, 15–19, 25, 29, 31–33, 39, 68, 97,
    98, 100–103, 106, 158, 161
  map, 125
  subgroup, 101, 105
  vector, 15, 24, 27, 31
triangle inequality, 1–5, 38, 45, 180
trichotomy, 177–179

ultraparallel lines, 48–51, 59, 61
UMP, see *universal mapping property*
unitary
  group, 153, 176
  matrix, 153, 158, 188–189
  representation, 175
universal mapping property, 118, 139,
    152

winding number, xv, 107, 130–137